Lecture Notes in Mathematics 2193

More information about this series at http://www.springer.com/series/304

Sergei Yu. Pilyugin • Kazuhiro Sakai

Shadowing and Hyperbolicity

Sergei Yu. Pilyugin
Faculty of Mathematics and Mechanics
St. Petersburg State University
St. Petersburg, Russia

Kazuhiro Sakai
Faculty of Education
Utsunomiya University
Utsunomiya, Japan

ISSN 0075-8434 ISSN 1617-9692 (electronic)
Lecture Notes in Mathematics
ISBN 978-3-319-65183-5 ISBN 978-3-319-65184-2 (eBook)
DOI 10.1007/978-3-319-65184-2

Library of Congress Control Number: 2017950671

Mathematics Subject Classification (2010): 37C50, 37D20, 37C20, 37D30

Printed on acid-free paper

This Springer imprint is published by Springer Nature
The registered company is Springer International Publishing AG
The registered company address is: Gewerbestrasse 11, 6330 Cham, Switzerland

To the memory of Dmitrii Viktorovich Anosov

Preface

The theory of shadowing of approximate trajectories (pseudotrajectories) in dynamical systems is now an important and rapidly developing branch of the modern global theory of dynamical systems.

The notion of a pseudotrajectory goes back to Birkhoff [9]. The real development of the shadowing theory started after the classical results of Anosov [4] and Bowen [12]. The main results obtained in the 20th century were reflected in the monographs [64] and [56]; one can find a survey of recent results in [70].

In fact, the modern shadowing theory has been developing on the powerful basis of the theory of structural stability, one of the main parts of the global theory of dynamical systems in the second half of the 20th century.

Undoubtedly, the notions of hyperbolicity and transversality, which are the key notions of the theory of structural stability, have become the basic notions of the shadowing theory as well.

This book is devoted to several recent results relating various shadowing properties to structural stability. It was understood in the 1970s that structural stability implies shadowing. In fact, both monographs [64] and [56] were mostly devoted to various proofs of the general statement: "Hyperbolicity (structural stability) implies shadowing." At the same time, simple examples show that shadowing is not equivalent to structural stability.

Nevertheless, all such examples are, in a sense, "degenerate," and it is natural to assume that in "nondegenerate" cases, shadowing and structural stability are equivalent. In a precise form, this assumption was formulated as a conjecture by Abdenur and Diaz in [1]; in fact, they conjectured that a C^1-generic diffeomorphism with shadowing is structurally stable and proved this conjecture for so-called tame diffeomorphisms.

Two more possible approaches to the problem of equivalence of shadowing and structural stability are related to the passage to C^1 interiors of the sets of dynamical systems having various shadowing properties or to study of special shadowing properties (such as Lipschitz and Hölder).

This book is devoted to the main results related to the above-mentioned two approaches, which, in our opinion, may be of interest for specialists in the global theory of dynamical systems.

One of our main goals was to give either complete proofs of some results which are not easily available (for example, the Pliss theorem used in Chap. 2 was only published in Russian long ago in proceedings of a Kiev conference and was never reproduced) or detailed expositions of some heavy proofs (for example, we do not reproduce the Tikhomirov's paper devoted to Hölder shadowing but instead explain in detail the not so heavy one-dimensional case to help the reader to understand the original proof).

Let us describe the contents of the book.

The book consists of four chapters.

Chapter 1 is preliminary. In this chapter, we define pseudotrajectories and various shadowing properties for dynamical systems with discrete and continuous time (Sects. 1.1 and 1.2), study the notion of chain transitivity (Sect. 1.1), describe hyperbolicity, Ω-stability, and structural stability (Sect. 1.3), and prove a lemma on finite Lipschitz shadowing in a neighborhood of a hyperbolic set (Sect. 1.4).

In Chap. 2, we give either complete proofs or schemes of proof of the following main results:

- If a diffeomorphism f of a smooth closed manifold has the Lipschitz shadowing property, then f is structurally stable (Theorem 2.3.1);
- a diffeomorphism f has the Lipschitz periodic shadowing property if and only if f is Ω-stable (Theorem 2.4.1);
- if a diffeomorphism f of class C^2 has the Hölder shadowing property on finite intervals with constants $\mathcal{L}, C, d_0, \theta, \omega$, where $\theta \in (1/2, 1)$ and $\theta + \omega > 1$, then f is structurally stable (Theorem 2.5.1);
- there exists a homeomorphism of the interval that has the Lipschitz shadowing property and a nonisolated fixed point (Theorem 2.6.1);
- if a vector field X has the Lipschitz shadowing property, then X is structurally stable (Theorem 2.7.1).

Since Theorem 2.3.1 is one of the basic results related to the study of Lipschitz shadowing property for diffeomorphisms, we include in the book complete versions of the main ingredients of its proof: in Sect. 2.1, we prove Maizel' and Pliss theorems relating the so-called Perron property of difference equations and hyperbolicity of sequences of linear automorphisms, Sect. 2.2 is devoted to the Mañé theorem characterizing structural stability in terms of the so-called analytic transversality condition (Theorem 1.3.7), and in Sect. 2.3, we reduce the proof of Theorem 2.3.1 to results of the previous two sections.

In Chap. 3, we study the structure of C^1 interiors of some basic sets of dynamical systems having various shadowing properties. We give either complete proofs or schemes of proof of the following main results:

- The C^1 interior of the set of diffeomorphisms having the standard shadowing property is a subset of the set of structurally stable diffeomorphisms (Theorem 3.1.1); this result combined with a well-known statement (structurally stable diffeomorphisms have the standard shadowing property) implies that the C^1 interior of the set of diffeomorphisms having the standard shadowing property coincides with the set of structurally stable diffeomorphisms;

- the C^1 interior of the set of vector fields having the oriented shadowing property minus some special set $\mathscr{B}$ of vector fields (consisting of vector fields that have a couple of rest points connected by a trajectory of nontransverse intersection of their stable and unstable manifolds; of course, such vector fields are not structurally stable) is a subset of the set of structurally stable vector fields (Theorem 3.3.1); similarly to the case of diffeomorphisms, this result combined with a well-known statement (structurally stable vector fields have the shadowing property) implies that the C^1 interior of the set of vector fields having the oriented shadowing property minus the set $\mathscr{B}$ coincides with the set of structurally stable vector fields;
- the C^1 interior of the set of vector fields having the oriented shadowing property contains vector fields that are not structurally stable (Theorem 3.4.1).

The structure of the chapter is as follows.

Section 3.1 is devoted to the proof of Theorem 3.1.1. Our proof of Theorem 3.1.1 is based on reduction to the theorem stating that the C^1 interior of the set of Kupka–Smale diffeomorphisms coincides with the set of structurally stable diffeomorphisms.

We give a detailed proof of the fact that any periodic point of a diffeomorphism in the C^1 interior of the set of diffeomorphisms having the standard shadowing property is hyperbolic. Concerning the proof of transversality of stable and unstable manifolds of periodic points of such a diffeomorphism, we refer the reader to Sect. 3.3 where a similar statement is proved in a more complicated case of flows on manifolds.

One of the necessary and sufficient conditions of structural stability of a diffeomorphism is Axiom A. In Sect. 3.2, we give an independent proof of the following statement, Theorem 3.2.1: If a diffeomorphism f belongs to the C^1 interior of the set of diffeomorphisms having the standard shadowing property, then f satisfies Axiom A. Our proof uses neither Mañé's ergodic closing lemma [42] nor the techniques of creating homoclinic orbits developed in [44]. Instead, we refer to a sifting type lemma of Wen–Gan–Wen [109] influenced by Liao's work and apply it to Liao's closing lemma.

Sections 3.3 and 3.4 are devoted to the study of the C^1 interior of the set of vector fields having the oriented shadowing property. We introduce the above-mentioned class $\mathscr{B}$ and prove Theorem 3.3.1.

In Sect. 3.4, we show that the C^1 interior of the set of vector fields having the oriented shadowing property contains vector fields belonging to $\mathscr{B}$. The complete description of the corresponding example given in [69] is quite complicated, and we describe a "model" suggested in [100].

In Chap. 4, we study relations between the shadowing property of diffeomorphisms on their chain transitive sets and the hyperbolicity of such sets.

We prove the following two main results:

- Let Λ be a closed invariant set of $f \in \mathrm{Diff}^1(M)$. Then $f|_\Lambda$ is chain transitive and C^1-stably shadowing in a neighborhood of Λ if and only if Λ is a hyperbolic basic set (Theorem 4.2.1);
- there is a residual set $\mathscr{R} \subset \mathrm{Diff}^1(M)$ such that if $f \in \mathscr{R}$ and Λ is a locally maximal chain transitive set of f, then Λ is hyperbolic if and only if $f|_\Lambda$ is shadowing (Theorem 4.3.1).

The structure of the chapter is as follows.

In Sect. 4.1, we discuss several examples of chain transitive sets. Section 4.2 is devoted to the proof of Theorem 4.2.1. In Sect. 4.3, we prove Theorem 4.3.1.

Each section of the book contains Historical Remarks.

The authors are really grateful to A. A. Rodionova who put a lot of time and effort into preparation of this book for publication.

St. Petersburg, Russia — Sergei Yu. Pilyugin
Utsunomiya, Japan — Kazuhiro Sakai
June 2017

Acknowledgements

The first author was supported by the Chebyshev Laboratory, Faculty of Mathematics and Mechanics, St. Petersburg State University, and by the RFBR (grant 15-01-03797a).

The second author was supported by JSPS KAKENHI Grant Number 16K05167.

Contents

1 **Main Definitions and Basic Results** 1
1.1 Pseudotrajectories and Shadowing in Dynamical Systems with Discrete Time: Chain Transitive Sets 1
1.2 Pseudotrajectories and Shadowing in Dynamical Systems with Continuous Time 9
1.3 Hyperbolicity, Ω-Stability, Structural Stability, Dominated Splittings 12
1.4 Hyperbolic Shadowing 26

2 **Lipschitz and Hölder Shadowing and Structural Stability** 37
2.1 Maizel' and Pliss Theorems 38
2.2 Mañé Theorem 51
2.3 Diffeomorphisms with Lipschitz Shadowing 67
2.4 Lipschitz Periodic Shadowing for Diffeomorphisms 75
2.5 Hölder Shadowing for Diffeomorphisms 90
2.6 A Homeomorphism with Lipschitz Shadowing and a Nonisolated Fixed Point 102
2.7 Lipschitz Shadowing Implies Structural Stability: The Case of a Vector Field 109
2.7.1 Discrete Lipschitz Shadowing for Flows 110
2.7.2 Rest Points 115
2.7.3 Hyperbolicity of the Chain Recurrent Set 118
2.7.4 Transversality of Stable and Unstable Manifolds 119

3 **C^1 Interiors of Sets of Systems with Various Shadowing Properties** 125
3.1 C^1 Interior of SSP_D 126
3.2 Diffeomorphisms in $Int^1(SSP_D)$ Satisfy Axiom A 133
3.3 Vector Fields in $Int^1(OrientSP_F \setminus \mathscr{B})$ 155
3.4 Vector Fields of the Class $\mathscr{B}$ 172

4 Chain Transitive Sets and Shadowing 181
4.1 Examples of Chain Transitive Sets (Homoclinic Classes) 181
4.1.1 Chain Transitive Sets Without Periodic Points 183
4.1.2 Hyperbolic Horseshoes 183
4.1.3 Horseshoe with a Homoclinic Tangency 184
4.1.4 Critical Saddle-Node Horseshoe 185
4.2 C^1-Stably Shadowing Chain Transitive Sets 187
4.2.1 Preliminaries 188
4.2.2 Construction of the Dominated Splitting and Its Extension 190
4.2.3 Proof of Theorem 4.2.1 194
4.2.4 Proof of Corollary 4.2.1 201
4.3 Chain Transitive Sets with Shadowing for Generic Diffeomorphisms 203

References 209

Index 215

Chapter 1
Main Definitions and Basic Results

In this preliminary chapter, we define pseudotrajectories and various shadowing properties for dynamical systems with discrete and continuous time (Sects. 1.1 and 1.2), study the notion of chain transitivity (Sect. 1.1), describe hyperbolicity, Ω-stability, and structural stability (Sect. 1.3), and prove a lemma on finite Lipschitz shadowing in a neighborhood of a hyperbolic set (Sect. 1.4).

1.1 Pseudotrajectories and Shadowing in Dynamical Systems with Discrete Time: Chain Transitive Sets

Consider a metric space (M, dist). Everywhere below (if otherwise is not stated), we denote by $N(a, x)$ and $N(a, A)$ the open a-neighborhoods of a point $x \in M$ and a set $A \subset M$, respectively. For a set $A \subset M$, $\text{Int}(A)$, $\text{Cl}(A)$, and ∂A denote the interior, closure, and boundary of A, respectively.

Let f be a homeomorphism of the metric space M. As usual, we identify the homeomorphism f with the dynamical system with discrete time generated by f on M.

We denote by

$$O(x,f) = \{f^k(x) : k \in \mathbb{Z}\}$$

the trajectory (orbit) of a point $x \in M$ in the dynamical system f.

We also consider positive and negative semitrajectories of a point x,

$$O^+(x,f) = \{f^k(x) : k \geq 0\} \text{ and } O^-(x,f) = \{f^k(x) : k \leq 0\}.$$

S.Yu. Pilyugin, K. Sakai, *Shadowing and Hyperbolicity*, Lecture Notes in Mathematics 2193, DOI 10.1007/978-3-319-65184-2_1

Similar notation is used for trajectories of sets;

$$O(A,f) = \{f^k(A) : k \in \mathbb{Z}\}$$

is the trajectory of a set $A \subset M$ in the dynamical system f, etc.

We denote by $\mathrm{Per}(f)$ the set of periodic points of f.

Remark 1.1.1 We give the main definitions in this section for the most general case of dynamical system with discrete time generated by homeomorphisms; in fact, the main results of this book are related to smooth dynamical systems – either to systems with discrete time generated by diffeomorphisms or to systems with continuous time (flows) generated by smooth vector fields on manifolds.

If M is a smooth closed (i.e., compact and boundaryless) manifold with Riemannian metric dist, we denote by TM the tangent bundle of M and by T_xM the tangent space of M at a point x, respectively. For a vector $v \in T_xM$, $|v|$ is its norm induced by the metric dist.

If f is a diffeomorphism of a smooth manifold M, we denote by

$$Df(x) : T_xM \to T_{f(x)}M$$

its derivative at a point $x \in M$.

Let us give the main definition in the case of a homeomorphism of a metric space (M, dist).

Definition 1.1.1 Fix a $d > 0$. A sequence

$$\xi = \{x_k \in M : k \in \mathbb{Z}\} \tag{1.1}$$

is called a *d-pseudotrajectory* of the dynamical system f if the following inequalities hold:

$$\mathrm{dist}(x_{k+1}, f(x_k)) < d, \quad k \in \mathbb{Z}. \tag{1.2}$$

Sometimes, d-pseudotrajectories are called *d-orbits*.

The basic property of dynamical systems related to the notion of a pseudotrajectory is called shadowing (or tracing).

Definition 1.1.2 We say that a dynamical system f has the *shadowing property* if for any $\varepsilon > 0$ we can find a $d > 0$ such that for any d-pseudotrajectory ξ of f there exists a point $p \in M$ such that

$$\mathrm{dist}\left(x_k, f^k(p)\right) < \varepsilon, \quad k \in \mathbb{Z}. \tag{1.3}$$

In this case, we say that the pseudotrajectory ξ is *ε-shadowed* by the exact trajectory of the point p, and the trajectory $O(p,f)$ is called the *shadowing trajectory*.

Sometimes, this property is called the *standard shadowing property* or the POTP (*pseudoorbit tracing property*, see [5] and [6]).

In addition to infinite pseudotrajectories, we consider also finite pseudotrajectories, i.e., sets of points

$$\xi = \{x_k \in M : \, l \leq k \leq m\}$$

such that analogs of inequalities (1.2) hold for $l \leq k \leq m-1$.

The corresponding shadowing property called *finite shadowing property* means that for any $\varepsilon > 0$ we can find a $d > 0$ such that for any finite d-pseudotrajectory ξ of f as above there exists a point $p \in M$ such that analogs of inequalities (1.3) hold for $l \leq k \leq m-1$. Here it is important to emphasize that d depends on ε and does not depend on the number $m - l$.

In what follows, it will be convenient for us to introduce special notation for sets of dynamical systems having some shadowing properties. Let us denote by SSP_D the set of systems with discrete time having the standard shadowing property (of course, any time, using a notation of that kind, we will indicate the phase space and the class of smoothness of the considered dynamical systems).

In this book, we also consider several modifications of the standard shadowing property.

The first of these modifications is a property that is weaker than the standard shadowing property. First let us recall the definition of the Hausdorff metric.

Denote by $\mathscr{C}(M)$ the set of all nonempty compact subsets of M. Let $x \in M$ and $K \in \mathscr{C}(M)$; set

$$\mathrm{dist}(x, K) = \min_{y \in K} \mathrm{dist}(x, y).$$

The Hausdorff metric dist_H on $\mathscr{C}(X)$ is defined as follows:

$$\mathrm{dist}_H(A, B) = \max\left(\max_{a \in A} \mathrm{dist}(a, B), \max_{b \in B} \mathrm{dist}(b, A) \right)$$

for $A, B \in \mathscr{C}(X)$.

The next result which we use below is well known (see p. 47 of [32]).

Lemma 1.1.1 *If the space M is compact, then $(\mathscr{C}(M), \mathrm{dist}_H)$ is a compact metric space.*

Definition 1.1.3 We say that a dynamical system f has the *orbital shadowing property* if for any $\varepsilon > 0$ we can find a $d > 0$ such that for any d-pseudotrajectory ξ of f there exists a point $p \in M$ such that

$$\mathrm{dist}_H(\mathrm{Cl}(\xi), \mathrm{Cl}(O(p, f))) < \varepsilon. \tag{1.4}$$

We denote by OSP_D the set of systems with discrete time having the orbital shadowing property.

One more shadowing property is defined below.

Definition 1.1.4 We say that f has the *Lipschitz shadowing property* if there exist $\mathscr{L}, d_0 > 0$ such that for any d-pseudotrajectory $\{x_k\}$ with $d \leq d_0$ there exists an exact trajectory $\{f^k(p)\}$ satisfying the inequalities

$$\operatorname{dist}\left(x_k, f^k(p)\right) \leq \mathscr{L}d, \quad k \in \mathbb{Z}. \tag{1.5}$$

One can define the *finite Lipschitz shadowing property* similarly to the finite shadowing property (we leave details to the reader).

Let us denote by LSP_D the set of systems with discrete time having the Lipschitz shadowing property.

Obviously, the following inclusions hold:

$$\mathrm{LSP}_D \subset \mathrm{SSP}_D \subset \mathrm{OSP}_D \tag{1.6}$$

(of course, here we have in mind that we consider dynamical systems with the same phase spaces).

Simple examples show that all the inclusions in (1.6) are strict.

To show that $\mathrm{SSP}_D \setminus \mathrm{LSP}_D \neq \emptyset$, consider a North Pole – South Pole diffeomorphism f of the circle S^1 that has two fixed points, an asymptotically stable fixed point s and a completely unstable (i.e., asymptotically stable for f^{-1}) fixed point u and such that $f^k(x) \to s, k \to \infty$, for any $x \neq u$, and $f^k(x) \to u, k \to -\infty$, for any $x \neq s$. It is easy to show that such a diffeomorphism f has the standard shadowing property. Theorem 1.4.1 (1) implies that if the fixed points s and u are hyperbolic (in this case, f is structurally stable), then f has the Lipschitz shadowing property. At the same time, it is an easy exercise to show that f does not have the Lipschitz shadowing property if one of the fixed points s or u is not hyperbolic.

It is also an easy exercise to show that irrational rotation of the circle gives us an example of a diffeomorphism belonging to $\mathrm{OSP}_D \setminus \mathrm{SSP}_D$.

It is possible to study shadowing properties dealing with pseudotrajectories that are subjected to some additional restrictions. In this book, we consider the case of periodic pseudotrajectories.

Definition 1.1.5 We say that f has the *periodic shadowing property* if for any $\varepsilon > 0$ we can find a $d > 0$ such that for any periodic d-pseudotrajectory ξ of f there exists a periodic point p of f such that inequalities (1.3) hold.

Remark 1.1.2 Note that it is not assumed in the above definition that the periods of the pseudotrajectory ξ and periodic point p coincide.

Let us denote by PerSP_D the set of systems with discrete time having the periodic shadowing property.

Definition 1.1.6 We say that f has the *Lipschitz periodic shadowing property* if there exist positive constants $\mathscr{L}, d_0$ such that if $\xi = \{x_k\}$ is a periodic

d-pseudotrajectory with $d \leq d_0$, then there exists a periodic point p of f such that inequalities (1.5) hold.

Let us denote by LPerSP_D the set of systems with discrete time having the Lipschitz periodic shadowing property.

As was mentioned, we also consider pseudotrajectories defined not on $\mathbb{Z}$ but on some subsets of $\mathbb{Z}$. Such pseudotrajectories will appear, for example, in the study of the following property.

Definition 1.1.7 We say that f has the *Hölder shadowing property on finite intervals* with constants $\mathcal{L}, C, d_0, \theta, \omega > 0$ if for any d-pseudotrajectory

$$\xi = \{x_k : 0 \leq k \leq Cd^{-\omega}\}$$

of f with $d \leq d_0$ there exists a point p such that

$$\operatorname{dist}\left(x_k, f^k(p)\right) \leq \mathcal{L} d^{\theta}, \quad 0 \leq k \leq Cd^{-\omega}. \tag{1.7}$$

We denote by $\mathrm{FHSP}_D(\mathcal{L}, C, d_0, \theta, \omega)$ the set of systems with discrete time having the property formulated in Definition 1.1.7.

An important application of pseudotrajectories defined on subsets of $\mathbb{Z}$ is the theory of chain recurrence and chain transitivity.

The main tools in this theory are ε-chains (finite ε-trajectories joining points of the phase space; following tradition, we preserve this terminology and use ε instead of d in analogs of inequalities (1.2)).

Until the end of this section, we assume, in addition, that M is a compact metric space.

Let C be a subset of M and let $p, q \in C$.

Definition 1.1.8 For $\varepsilon > 0$, a sequence $\{x_0, x_1, \dots, x_m\}$ of points of the subset C is called an *ε-chain* in C of length $m + 1$ from p to q if $x_0 = p$, $x_m = q$, and $\operatorname{dist}(f(x_i), x_{i+1}) < \varepsilon$ for $0 \leq i < m$.

If there is an ε-chain in C from p to q, then we write $p \rightsquigarrow_C^{\varepsilon} q$.

Let us also write

$$p \leftrightsquigarrow_C^{\varepsilon} q \text{ if both } p \rightsquigarrow_C^{\varepsilon} q \text{ and } q \rightsquigarrow_C^{\varepsilon} p,$$

$$p \rightsquigarrow_C q \text{ if } p \rightsquigarrow_C^{\varepsilon} q \text{ for any } \varepsilon > 0,$$

$$p \leftrightsquigarrow_C q \text{ if } p \leftrightsquigarrow_C^{\varepsilon} q \text{ for any } \varepsilon > 0.$$

In the above notation, we omit C if $C = M$.

Definition 1.1.9 A point $x \in M$ is called a *chain recurrent point* if $x \leftrightsquigarrow x$.

Definition 1.1.10 The set

$$\mathscr{R}(f) = \{x \in M : x \leftrightsquigarrow x\}$$

of all chain recurrent points of f is called the *chain recurrent set of f*.

Definition 1.1.11 Two points x and y of M are called *chain equivalent* if $x \leftrightsquigarrow y$.

Note that if $x, y \in M$ and $x \leftrightsquigarrow y$, then $x, y \in \mathscr{R}(f)$.

Clearly, the chain equivalence is an equivalence relation on $\mathscr{R}(f)$.

Definition 1.1.12 Each equivalence class of the above equivalence relation is called a *chain recurrence class*.

We note that $\mathscr{R}(f)$ and chain recurrence classes are closed f-invariant sets (see Lemma 1.1.5 below).

Definition 1.1.13 We say that a closed f-invariant set Λ is *chain transitive* if $x \rightsquigarrow_{\Lambda} y$ for any $x, y \in \Lambda$.

A chain recurrence class $\mathscr{R}$ is called a *maximal chain transitive set* if the inclusion $\mathscr{R} \subset C$, where C is a chain transitive set, implies that $\mathscr{R} = C$.

The main statement which we prove in this section is the following proposition.

Proposition 1.1.1 *Any chain recurrence class is a maximal chain transitive set.*

The next convention will be frequently used in this section. For $\varepsilon > 0$, $\delta(\varepsilon)$ denotes a real number such that $0 < \delta(\varepsilon) < \varepsilon$ and the inequality $\mathrm{dist}(x, y) < \delta(\varepsilon)$ implies that $\mathrm{dist}(f(x), f(y)) < \epsilon$.

We prove a sequence of lemmas which we need.

Lemma 1.1.2 *The relation*

$$\mathscr{R}(\rightsquigarrow) = \{(x, y) \in M \times M : x \rightsquigarrow y\}$$

is closed in $M \times M$.

Proof Let a sequence $\{(x_i, y_i) : 1 \leq i < \infty\}$ in $\mathscr{R}(\rightsquigarrow)$ converge to $(x, y) \in M \times M$. We show that $(x, y) \in \mathscr{R}(\rightsquigarrow)$. For $\varepsilon > 0$, let $\delta = \delta(\varepsilon/3)$. Fix an index $i \geq 1$ such that $\max(\mathrm{dist}(x_i, x), \mathrm{dist}(y_i, y)) < \delta$. Since $x_i \rightsquigarrow y_i$, there is a δ-chain $\{z_0, \dots, z_m\}$ from x_i to y_i. Assume that $m = 1$. Then

$$\mathrm{dist}(f(x), y) \leq \mathrm{dist}(f(x), f(z_0)) + \mathrm{dist}(f(z_0), z_1) + \mathrm{dist}(z_1, y) <$$
$$< \varepsilon/3 + \delta + \delta < \varepsilon.$$

Thus, $x \rightsquigarrow^{\epsilon} y$. Next assume that $m \geq 2$. Then it is easy to see that

$$\{x, z_1, z_2, \dots, z_{m-1}, y\}$$

is an ε-chain from x to y. Hence, $x \rightsquigarrow^{\epsilon} y$ in any case. Since $\varepsilon > 0$ is arbitrary, $x \rightsquigarrow y$, and $(x, y) \in \mathscr{R}(\rightsquigarrow)$. □

The following statement is an obvious corollary of Lemma 1.1.2.

Lemma 1.1.3 *The relation*

$$\mathscr{R}(\leftrightsquigarrow) = \{(x, y) \in M \times M : x \leftrightsquigarrow y\}$$

is closed in $M \times M$.

Lemma 1.1.4

$$(f \times f)(\mathscr{R}(\rightsquigarrow)) \subset \mathscr{R}(\rightsquigarrow)$$

and

$$(f \times f)(\mathscr{R}(\leftrightsquigarrow)) \subset \mathscr{R}(\leftrightsquigarrow).$$

Proof It is enough to prove the first inclusion. Let $(x, y) \in \mathscr{R}(\rightsquigarrow)$; we show that $(f(x), f(y)) \in \mathscr{R}(\rightsquigarrow)$. Fix an $\varepsilon > 0$ and let $\delta = \delta(\varepsilon)$. Since $x \rightsquigarrow y$, there is a δ-chain $\{x_0, \dots, x_m\}$ from x to y. It is easy to see that $\{f(x_0), \dots, f(x_m)\}$ is an ε-chain from $f(x)$ to $f(y)$. Thus, $f(x) \rightsquigarrow f(y)$. Since $\epsilon > 0$ is arbitrary, $(f(x), f(y)) \in \mathscr{R}(\rightsquigarrow)$. □

Lemma 1.1.5 *The set* $\mathscr{R}(f)$ *and each chain recurrence class are closed* f*-invariant sets.*

Proof Let A be a chain recurrence class of f. It follows directly from Lemma 1.1.3 that both $\mathscr{R}(f)$ and A are closed. Since $\mathscr{R}(f)$ is a disjoint union of chain recurrence classes, it is enough to show that A is f-invariant.

Let $x \in A$. Then for each $n \geq 1$ there is a $(1/n)$-chain $\{x_0^n, \dots, x_{m_n}^n\}$ from x to itself. Put $y_n = x_{m_n-1}^n$, $n \geq 1$, and let y be one of the limit points of the sequence $\{y_n : n \geq 1\}$. It is easy to see that $x \rightsquigarrow y$. Since $\text{dist}(f(y), x) < 1/n$ for $n \geq 1$, we get the equality $f(y) = x$. Hence, $f(x) \rightsquigarrow f(y) = x$ by Lemma 1.1.4. Since $y \rightsquigarrow f(y) = x \rightsquigarrow f(x)$, we conclude that $x \leftrightsquigarrow y$ and $x \leftrightsquigarrow f(x)$. Thus, both y and $f(x)$ are chain recurrent points and belong to A. Since $x \in A$ is arbitrary, it follows that $f(A) \supset A \supset f(A)$, i.e., $f(A) = A$. □

Let, as above, $\mathscr{C}(M)$ be the set of all nonempty compact subsets of M with the Hausdorff metric dist_H (by Lemma 1.1.1, $(\mathscr{C}(M), \text{dist}_H)$ is a compact metric space).

Consider the map $\mathscr{C}(f) : \mathscr{C}(M) \to \mathscr{C}(M)$ defined by $\mathscr{C}(f)(A) = f(A)$ for $A \in \mathscr{C}(X)$. Clearly, this map is continuous.

Recall that a closed f-invariant subset A is chain transitive if $x \leftrightsquigarrow_A y$ for all $x, y \in A$.

Proof (of Proposition 1.1.1) Let A be a chain recurrence class.

By Lemma 1.1.5, A is closed and $f(A) = A$. We prove the proposition modifying the proof of the result of Robinson [84]. Let $x, y \in A$. For each integer $n \geq 1$, take a $(1/n)$-chain $C_n = \{x_0^n, \dots, x_{m_n}^n\}$ from x through y to x. In particular, $x, y \in C_n$. Since $C_n \in \mathscr{C}(M)$ for any n, there is a subsequence n_k such that $\lim_{k\to\infty} C_{n_k} = C$ for some $C \in \mathscr{C}(M)$. Note that $x, y \in C$. We show that $f(C) = C$. Since $x_0^n = x_{m_n}^n$, we see that $\text{dist}_H(f(C_n), C_n) < 1/n$. Thus,

$$\text{dist}_H(f(C), C) \le \text{dist}_H(f(C), f(C_{n_k})) +$$

$$+\text{dist}_H(f(C_{n_k}), C_{n_k}) + \text{dist}_H(C_{n_k}, C) \le$$

$$\le \text{dist}_H\,(f(C), f(C_{n_k})) + \frac{1}{n_k} + \text{dist}_H(C_{n_k}, C).$$

Letting $k \to \infty$, we conclude that $\text{dist}_H(f(C), C) = 0$, i.e., $f(C) = C$.

Next we show that C is chain transitive. Let $z, w \in C$, and fix any $\varepsilon > 0$. Let $\delta = \delta(\varepsilon/3)$ and take $n = n_k$ such that $1/n < \varepsilon/3$ and $\text{dist}_H(C, C_n) < \delta$. Then

$$C \subset \bigcup_{i=0}^{m_n} N(\delta, x_i^n)$$

(recall that $N(\delta, x) = \{ y \in M : \text{dist}(y, x) < \delta\}$).

Take i, j with $0 \le i, j \le m_n$ such that $\text{dist}(z, x_i^n) < \delta$ and $\text{dist}(w, x_j^n) < \delta$. Since $x_0^n = x_{m_n}^n$, there is a $(1/n)$-chain $\{ y_0, y_1, \dots, y_m\} \subset C$ from x_i^n to x_j^n. We now construct an ε-chain $\{z_0, z_1, \dots, z_m\}$ in C from z to w. For $0 < k < m$, take $z_k \in C$ such that $z_k \in N(\delta, y_k)$, and let $z_0 = z$, $z_m = w$. Since $\text{dist}(z_k, y_k) < \delta = \delta(\varepsilon/3)$, it follows that

$$\text{dist}(f(z_k), z_{k+1}) \le \text{dist}(f(z_k), f(y_k)) + \text{dist}(f(y_k), y_{k+1}) +$$

$$+\text{dist}(y_{k+1}, z_{k+1}) < \varepsilon/3 + \frac{1}{n} + \delta < \varepsilon$$

for each $0 \le k < m$. Thus, $\{z_0, \dots, z_m\}$ is an ε-chain in C from z to w. Since $\varepsilon > 0$ is arbitrary, $z \rightsquigarrow_C w$. Since $z, w \in C$ are arbitrary, $z \leftrightsquigarrow_C w$ for any $z, w \in C$, i.e., C is chain transitive. If we take $x = z$, then $x \leftrightsquigarrow_C w$ for all $w \in C$. Thus, $C \subset A$. Since $x \leftrightsquigarrow_C y$, we conclude that $x \leftrightsquigarrow_A y$. Hence, A is chain transitive, as desired. The maximality of A is obvious. □

It is easy to see that the following statement holds (we omit the proof).

Lemma 1.1.6 *For any $x \in M$, the omega-limit set $\omega_f(x)$ of x and alpha-limit set $\alpha_f(x)$ of x are chain transitive.*

Historical Remarks Pseudotrajectories of a special kind (called δ-chains) were considered by G. D. Birkhoff in his study of the last Poincaré geometric theorem [9].

The first basic results related to shadowing were obtained by D. V. Anosov and R. Bowen in [4] and [12] for hyperbolic sets of diffeomorphisms. It is easily seen that both Anosov's and Bowen's proofs, in fact, give Lipschitz shadowing in a neighborhood of a hyperbolic set of a diffeomorphism.

The orbital shadowing property was first considered in the joint paper [65] of the authors of this book and A. A. Rodionova.

The periodic and Lipschitz periodic shadowing were studied by A. V. Osipov, the first author of this book, and S. B. Tikhomirov in [50].

S. B. Tikhomirov studied the Hölder shadowing property on finite intervals in the paper [101].

C. Conley introduced the notion of chain recurrence in [14] and [15]. Most of the results of this section devoted to chain recurrence and chain transitivity, which were reformulated for discrete dynamical systems in Shimomura [93], can be found in [14] and [15] in the case of flows. As far as we know, chain transitive sets of discrete dynamical systems with the standard shadowing property were first considered in [93] from the view point of topological entropy.

1.2 Pseudotrajectories and Shadowing in Dynamical Systems with Continuous Time

Let M be a smooth closed manifold. Consider a C^1 vector field X on M and denote by ϕ the flow of X. We denote by

$$O(x, \phi) = \{\phi(t, x) : t \in \mathbb{R}\}$$

the trajectory of a point x in the flow ϕ; $O^+(x, \phi)$ and $O^-(x, \phi)$ are the positive and negative semitrajectories, respectively.

Definition 1.2.1 Fix a number $d > 0$. We say that a mapping $g : \mathbb{R} \to M$ (not necessarily continuous) is a *d-pseudotrajectory* (both for the field X and flow ϕ) if

$$\operatorname{dist}(g(\tau + t), \phi(t, g(\tau))) < d \quad \text{for} \quad \tau \in \mathbb{R},\ t \in [0, 1]. \tag{1.8}$$

Of course, one can also define finite pseudotrajectories defined not on $\mathbb{R}$ but on finite segments $[a, b]$. We leave details to the reader.

It is easy to understand that, defining shadowing properties in the case of flows, it is not reasonable to give a definition parallel to Definition 1.1.2 just replacing inequality (1.3) by an inequality of the form

$$\operatorname{dist}(g(t), \phi(t, p)) < \varepsilon, \quad t \in \mathbb{R}. \tag{1.9}$$

Indeed, consider the following simple example.

Example 1.2.1 Let M be the two-dimensional sphere S^2; consider in a coordinate neighborhood U homeomorphic to $\mathbb{R}^2$ a vector field X having an isolated closed trajectory γ parametrized by

$$\xi(t) = (\sin t, \cos t), \quad t \in \mathbb{R}.$$

Take a small $d > 0$ and let

$$g(t) = \xi(t + kd/2), \quad t \in [2\pi k, 2\pi(k+1)), \; k \in \mathbb{Z}.$$

Since $|X(x)| = 1$ at points of γ, it is easy to understand that g is a d-pseudotrajectory of X.

Assume that there exists a point p such that inequality (1.9) holds with $\varepsilon = \varepsilon(d) \to 0$ as $d \to 0$. Since the trajectory γ is isolated, this is possible (for ε small enough) only if $p \in \gamma$. In this case, there exists a θ such that

$$\phi(t, p) = \xi(t + \theta).$$

Note that

$$\phi(2\pi k, p) = \xi(2\pi k + \theta) = \xi(\theta), \quad k \in \mathbb{Z},$$

while the set of points

$$g(2\pi k) = \xi(2\pi k + kd/2) = \xi(kd/2)$$

is d-dense in γ. Hence, for any d small enough there exists a k such that the distance between $g(2\pi k)$ and $\xi(2\pi k + \theta) = \xi(\theta)$ is larger than $\pi/2$, which contradicts our assumption.

Clearly, a similar construction can be realized in any flow having an isolated closed trajectory, and the set of such flows is large enough.

To avoid problems of that kind, one has to change parametrization of the shadowing trajectories. We introduce the following notion.

Definition 1.2.2 A *reparametrization* is an increasing homeomorphism h of the line $\mathbb{R}$; we denote by Rep the set of all reparametrizations.

For $a > 0$, we denote

$$\mathrm{Rep}(a) = \left\{ h \in \mathrm{Rep} : \left| \frac{h(t) - h(s)}{t - s} - 1 \right| < a, \quad t, s \in \mathbb{R}, \; t \neq s \right\}.$$

Definition 1.2.3 We say that a vector field X has the *standard shadowing property* if for any $\varepsilon > 0$ we can find $d > 0$ such that for any d-pseudotrajectory $g(t)$ of X there exists a point $p \in M$ and a reparametrization $h \in \mathrm{Rep}(\varepsilon)$ such that

$$\mathrm{dist}(g(t), \phi(h(t), p)) < \varepsilon \quad \text{for} \quad t \in \mathbb{R}. \tag{1.10}$$

We denote by SSP_F the set of vector fields having the standard shadowing property.

Definition 1.2.4 We say that a vector field X has the *Lipschitz shadowing property* if there exist $d_0 > 0$ and $\mathscr{L} > 0$ such that for any d-pseudotrajectory $g(t)$ of X with

$d \leq d_0$ there exists a point $p \in M$ and a reparametrization $h \in \text{Rep}(\mathcal{L}d)$ such that

$$\text{dist}(g(t), \phi(h(t), p)) \leq \mathcal{L}d \quad \text{for} \quad t \in \mathbb{R}. \tag{1.11}$$

We denote by LSP_F the set of vector fields having the Lipschitz shadowing property.

Definition 1.2.5 We say that a vector field X has the *oriented shadowing property* if for any $\varepsilon > 0$ we can find $d > 0$ such that for any d-pseudotrajectory $g(t)$ of X there exists a point $p \in M$ and a reparametrization $h \in \text{Rep}$ such that inequalities (1.10) hold (we emphasize that in this case, it is not assumed that the reparametrization h is close to identity).

We denote by OrientSP_F the set of vector fields having the oriented shadowing property.

Definition 1.2.6 We say that a vector field X has the *orbital shadowing property* if for any $\varepsilon > 0$ we can find $d > 0$ such that for any d-pseudotrajectory $g(t)$ of X there exists a point $p \in M$ such that

$$\text{dist}_H(\text{Cl}(\{g(t) : t \in \mathbb{R}\}), \text{Cl}(O(p, \phi))) < \varepsilon.$$

We denote by OrbitSP_F the set of vector fields having the orbital shadowing property.

Obviously, the following inclusions hold:

$$\text{OrbitSP}_F \supset \text{OrientSP}_F \supset \text{SSP}_F \supset \text{LSP}_F$$

(of course, here we have in mind that we consider vector fields on the same manifold).

It is easy to show that

$$\text{SSP}_F \setminus \text{LSP}_F \neq \emptyset.$$

It was recently shown by Tikhomirov [100] that

$$\text{OrientSP}_F \setminus \text{SSP}_F \neq \emptyset$$

(this solved the old problem posed by M. Komuro in [29]).

Historical Remarks Let us note that the standard shadowing property of vector fields (and their flows) is equivalent to the strong pseudo orbit tracing property (POTP) in the sense of M. Komuro [29] and [30]; the oriented shadowing property was called the normal POTP by M. Komuro [29] and the POTP for flows by R. F. Thomas in [102].

1.3 Hyperbolicity, Ω-Stability, Structural Stability, Dominated Splittings

Let us shortly recall the definitions of basic notions of the theory of structural stability of dynamical systems which we use in this book.

Let M be a smooth closed manifold and let f be a diffeomorphism of M of class C^1.

Definition 1.3.1 We say that a set $I \subset M$ is a *hyperbolic set* of a diffeomorphism f if the following conditions hold:

(HSD1) the set I is compact and f-invariant;
(HSD2) there exist numbers $C > 0$ and $\lambda \in (0, 1)$ and linear subspaces $S(p)$ and $U(p)$ of the tangent space T_pM defined for any point $p \in I$ such that
(HSD2.1) $S(p) \oplus U(p) = T_pM$;
(HSD2.2) $Df(p)S(p) = S(f(p))$ and $Df(p)U(p) = U(f(p))$;
(HSD2.3) if $v \in S(p)$, then $|Df^k(p)v| \leq C\lambda^k|v|$ for $k \geq 0$;
(HSD2.4) if $v \in U(p)$, then $|Df^k(p)v| \leq C\lambda^{-k}|v|$ for $k \leq 0$.

The numbers $C > 0$ and $\lambda \in (0, 1)$ are usually called *hyperbolicity constants* of the set I; the families $S(p)$ and $U(p)$ are called the *hyperbolic structure* on I.

The main objects related to a hyperbolic set I are stable and unstable manifolds of its points.

Definition 1.3.2 The *stable* and *unstable manifolds* of a point $p \in I$ are the sets defined by the equalities

$$W^s(p) = \left\{x \in M : \text{dist}\left(f^k(x), f^k(p)\right) \to 0,\ k \to \infty\right\}$$

and

$$W^u(p) = \left\{x \in M : \text{dist}\left(f^k(x), f^k(p)\right) \to 0,\ k \to -\infty\right\},$$

respectively.

The classical stable manifold theorem (see, for example, [27, 108]) states that if p is a point of a hyperbolic set I as above and $\sigma(p) = \dim S(p)$, then $W^s(p)$ is the image of the Euclidean space $\mathbb{R}^{\sigma(p)}$ under a C^1 immersion α_p^s; this means that the map

$$\alpha_p^s : \mathbb{R}^{\sigma(p)} \to W^s(p)$$

is one-to-one and that

$$\text{rank} D\alpha_p^s(x) = \sigma(p), \quad x \in \mathbb{R}^{\sigma(p)}.$$

In addition, $\alpha_p^s(0) = p$ and

$$T_p W^s(p) = S(p).$$

A similar statement (with $\sigma(p) = \dim U(p)$) is valid for $W^U(p)$.

One more classical definition which we need is the definition of the nonwandering set of a diffeomorphism f.

Definition 1.3.3 A point x is called a *nonwandering* point of f if for any neighborhood U of x and for any number N there exists a number n, $|n| > N$, such that $f^n(U) \cap U \neq \emptyset$. We denote by $\Omega(f)$ the set of nonwandering points of f (sometimes, the set $\Omega(f)$ is called the *nonwandering set* of f).

It is not difficult to show that the set $\Omega(f)$ is nonempty, compact, and f-invariant (see, for example, [71]).

Now we recall the two basic definitions of the theory of structural stability of diffeomorphisms, the definitions of Ω-stability and structural stability.

Let us start with the definition of the C^1 *topology* on the space of diffeomorphisms of a smooth closed manifold M.

First we define a C^0 metric ρ_0 on the space of homeomorphisms of a compact metric space.

Let (M, dist) be a compact metric space. If f and g are two homeomorphisms of the space M, we set

$$\rho_0(f, g) = \max_{x \in M} \max \left(\text{dist}(f(x), g(x)), \text{dist}(f^{-1}(x), g^{-1}(x)) \right). \tag{1.12}$$

It is easy to show that ρ_0 is a metric on the space of homeomorphisms of the space M.

We denote by $H(M)$ the space of homeomorphisms of the space M with the metric ρ_0; the topology induced by the metric ρ_0 is called the C^0 *topology*.

It is not difficult to show that the metric space $H(M)$ is complete (see, for example, [71]). At the same time, if we consider the topology on the space of homeomorphisms induced by the standard uniform metric

$$\max_{x \in M} \text{dist}(f(x), g(x)), \tag{1.13}$$

then the resulting space is not necessarily complete (see [71]).

Let now M be a smooth closed n-dimensional manifold. To introduce the C^1 topology on the space of diffeomorphisms of M, we assume that M is a submanifold of the Euclidean space $\mathbb{R}^N$ (a different, equivalent, approach to definition of the C^1 topology based on local coordinates is described in [60]).

No generality is lost assuming that M is a submanifold of a Euclidean space since, by the classical Whitney theorem, any smooth closed manifold can be embedded into a Euclidean space of appropriate dimension.

If M is a submanifold of $\mathbb{R}^N$, for any point $x \in M$ we can identify the tangent space $T_x M$ of M at x with a linear subspace of $\mathbb{R}^N$. Consider the metric dist on M

induced by the Euclidean metric of the space $\mathbb{R}^N$. For a vector $v \in T_xM$ we denote by $|v|$ its norm as the norm in the space $\mathbb{R}^N$.

Let f and g be two diffeomorphisms of the manifold M. Define the value $\rho_0(f, g)$ by the same formula (1.12) as for homeomorphisms of a compact metric space.

Take a point x of the manifold M and a vector v from the tangent space T_xM. We consider the vectors $Df(x)v \in T_{f(x)}M$ and $Dg(x)v \in T_{g(x)}M$ as vectors of the same Euclidean space $\mathbb{R}^N$. Hence, the following values are defined: $|Df(x)v - Dg(x)v|$ and

$$\|Df(x) - Dg(x)\| = \max_{v \in T_xM, |v|=1} |Df(x)v - Dg(x)v|.$$

Introduce the number

$$\rho_1(f, g) = \rho_0(f, g) + \max_{x \in M} \max \left(\|Df(x) - Dg(x)\|, \|Df^{-1}(x) - Dg^{-1}(x)\| \right).$$

Clearly, ρ_1 is a metric on the space of diffeomorphisms of the manifold M. We denote by $\mathrm{Diff}^1(M)$ the space of diffeomorphisms of M with the metric ρ_1; the topology induced by the metric ρ_1 is called the C^1 *topology*.

The standard reasoning shows that the topology on $\mathrm{Diff}^1(M)$ does not depend on the embedding of M into a Euclidean space and that $(\mathrm{Diff}^1(M), \rho_1)$ is a complete metric space.

Remark 1.3.1 To explain why it is reasonable to include the term $\|Df^{-1}(x) - Dg^{-1}(x)\|$ in the definition of the C^1 topology on the space of diffeomorphisms, let us consider the following example.

Let $M = S^1$ with coordinate $x \in [0, 1)$, fix a small $t \geq 0$ and define a mapping

$$f_t : S^1 \to S^1$$

by the formula

$$f_t(x) = tx + x^3 + h_t(x),$$

where h_t is of class C^1 in x,

$$h_t(x) = 0, \quad x \leq 1/3,$$

and

$$h_t(x) = 3x(1 - x) - t, \quad x \geq 2/3.$$

Then

$$f_t(x) = tx + x^3, \quad x \leq 1/3,$$

and

$$f_t(x) = 1 + t(x-1) + (x-1)^3, \quad x \geq 2/3.$$

Clearly, one can construct h_t so that

$$f_t'(x) > 0$$

for small $t > 0$ and for all x (thus, any f_t with such t is a diffeomorphism of S^1) and

$$\sup_{0 \leq x < 1} \left(|h_t(x) - h_\tau(x)| + |h_t'(x) - h_\tau'(x)| \right) \to 0, \quad t, \tau \to 0. \tag{1.14}$$

It follows from (1.14) that the family $\{f_t\}$ is a Cauchy sequence as $t \to 0$ with respect to the metric

$$\rho(f, g) = \sup_{0 \leq x < 1} \left(|f(x) - g(x)| + |f'(x) - g'(x)| \right)$$

but, clearly, its limit as $t \to 0$ is not a diffeomorphism of S^1.

Thus, the space of diffeomorphisms of S^1 with the metric ρ is not complete.

In what follows, if A is a subset of $\mathrm{Diff}^1(M)$, then $\mathrm{Int}^1(A)$ denotes the interior of A in $\mathrm{Diff}^1(M)$.

Definition 1.3.4 A diffeomorphism f is called *structurally stable* if there exists a neighborhood W of the diffeomorphism f in $\mathrm{Diff}^1(M)$ such that any diffeomorphism $g \in W$ is topologically conjugate to f (i.e., there exists a homeomorphism $h : M \to M$ such that $h \circ f = g \circ h$).

We denote by $\mathcal{S}_D(M)$ the set of structurally stable diffeomorphisms in $\mathrm{Diff}^1(M)$. We agree to write Diff^1 and $\mathcal{S}_D$ instead of $\mathrm{Diff}^1(M)$ and $\mathcal{S}_D(M)$, respectively, if it is not important for us to indicate the manifold M (as in the remark below).

Remark 1.3.2 Clearly, the set $\mathcal{S}_D$ is open in Diff^1.

Definition 1.3.5 A diffeomorphism f is called *Ω-stable* if there exists a neighborhood W of the diffeomorphism f in $\mathrm{Diff}^1(M)$ such that for any diffeomorphism $g \in W$ there exists a homeomorphism $h : \Omega(f) \to \Omega(g)$ such that

$$h \circ f|_{\Omega(f)} = g \circ h|_{\Omega(f)}.$$

We denote by $\Omega\mathcal{S}_D(M)$ (or simply $\Omega\mathcal{S}_D$) the set of Ω-stable diffeomorphisms. The following statements are also obvious.

Remark 1.3.3

(1) The set $\Omega\mathcal{S}_D$ is open in Diff^1.
(2) $\mathcal{S}_D \subset \Omega\mathcal{S}_D$.

Now we pass to characterization of Ω-stability and structural stability.

S. Smale introduced the following condition.

Axiom A

(AAa) The nonwandering set $\Omega(f)$ is hyperbolic.
(AAb) Periodic points of f are dense in $\Omega(f)$.

This condition played a very important role in the development of the theory of structural stability. First we describe the structure of the nonwandering set of a diffeomorphism that satisfies Axiom A. Smale proved the following statement.

Theorem 1.3.1 (Spectral Decomposition Theorem) *If a diffeomorphism f satisfies Axiom A, then its nonwandering set can be represented in the form*

$$\Omega(f) = \Omega_1 \cup \cdots \cup \Omega_m, \tag{1.15}$$

where the Ω_i are disjoint, compact, invariant sets such that each of these sets contains a dense positive semitrajectory. Representation of the form (1.15) is unique.

The sets Ω_i in representation (1.15) are called *basic*.

We can define analogs of stable and unstable manifolds for basic sets Ω_i:

$$W^s(\Omega_i) = \left\{x \in M : \ \mathrm{dist}\left(f^k(x), \Omega_i\right) \to 0, \quad k \to \infty\right\}$$

and

$$W^u(\Omega_i) = \left\{x \in M : \ \mathrm{dist}\left(f^k(x), \Omega_i\right) \to 0, \quad k \to -\infty\right\}.$$

The following statement holds (one can find a proof, for example, in [60]).

Theorem 1.3.2 *If a diffeomorphism f satisfies Axiom A, then*

$$M = \bigcup_{i=1}^{m} W^s(\Omega_i) = \bigcup_{i=1}^{m} W^u(\Omega_i). \tag{1.16}$$

Thus, any trajectory $f^k(x)$ of a diffeomorphism that satisfies Axiom A tends to a basic set as $|k| \to \infty$.

Now we give definitions which we need to formulate necessary and sufficient conditions of Ω-stability and structural stability of diffeomorphisms.

Let Ω_i and Ω_j be two (not necessarily different) basic sets of a diffeomorphism that satisfies Axiom A. We write $\Omega_i \to \Omega_j$ if there is a point $x \notin \Omega(f)$ such that

$$f^{-k}(x) \to \Omega_i \text{ and } f^k(x) \to \Omega_j, \quad k \to \infty.$$

Definition 1.3.6 We say that a diffeomorphism f has a 1-*cycle* if there exists a basic set Ω_i such that $\Omega_i \to \Omega_i$.

We say that a diffeomorphism f has a *k-cycle*, $k > 1$, if there exist k different basic sets $\Omega_{i_1}, \ldots, \Omega_{i_k}$ such that

$$\Omega_{i_1} \to \cdots \to \Omega_{i_k} \to \Omega_{i_1}.$$

We say that a diffeomorphism satisfies the *no cycle condition* if it does not have k-cycles with $k \geq 1$.

Theorem 1.3.3 *A diffeomorphism f is Ω-stable if and only if f satisfies Axiom A and the no cycle condition.*

Definition 1.3.7 Let f be a diffeomorphism satisfying Axiom A. We say that f satisfies the *geometric strong transversality condition* if stable and unstable manifolds of nonwandering points are transverse, i.e., if $p, q \in \Omega(f)$ and $x \in W^u(p) \cap W^s(q)$, then

$$T_x W^u(p) + T_x W^s(q) = T_x M. \tag{1.17}$$

Remark 1.3.4 Usually, the condition introduced in Definition 1.3.7 is called the strong transversality condition; we add the term *geometric* to distinguish this condition and the analytic strong transversality condition introduced below, in Definition 1.3.11.

Theorem 1.3.4 *A diffeomorphism f is structurally stable if and only if f satisfies Axiom A and the geometric strong transversality condition.*

Theorems 1.3.3 and 1.3.4 are classical basic results of the theory of structural stability. Nevertheless, sometimes it is more convenient to use different statements which characterize Ω-stability and structural stability (as we do in this book). Let us formulate some of them.

Recall that $\text{Per}(f)$ denotes the set of periodic points of a diffeomorphism f.

Definition 1.3.8 A periodic point p is called *hyperbolic* if its trajectory $O(p, f)$ is a hyperbolic set. It is easy to see that if p is a periodic point of period m, then p is hyperbolic if and only if the derivative $Df^m(p)$ does not have eigenvalues λ with $|\lambda| = 1$.

Denote by HP_D the set of diffeomorphisms f such that any periodic point of f is hyperbolic.

Theorem 1.3.5 *The sets $\text{Int}^1(\text{HP}_D)$ and $\Omega\mathcal{S}_D$ coincide.*

Sometimes, the set $\text{Int}^1(\text{HP}_D)$ is denoted by $\mathcal{F}$ and its elements are called *star systems*.

Remark 1.3.5 It follows from Theorem 1.3.5 that to establish the Ω-stability of a diffeomorphism f, it is enough to show that f and its C^1-small perturbations do not have nonhyperbolic periodic points.

Definition 1.3.9 A diffeomorphism $f \in \mathrm{HP}_D$ is called *Kupka–Smale* if stable and unstable manifolds of its periodic points are transverse. We denote by KS_D the set of Kupka–Smale diffeomorphisms.

Definition 1.3.10 A subset A of a topological space X is called *residual* if A contains the intersection of a countable family of open and dense subsets of X. A property P of elements of X is called *generic* if the set

$$\{x \in X : x \text{ satisfies } P\}$$

is residual.

Theorem 1.3.6

(1) The set KS_D is residual in $Diff^1$.
(2) The sets $Int^1(KS_D)$ and $\mathcal{S}_D$ coincide.

Remark 1.3.6 It follows from the second statement of Theorem 1.3.6 that to establish that a diffeomorphism f is structurally stable, it is enough to show that f has a C^1 neighborhood belonging to KS_D.

One more way of proving that a diffeomorphism is structurally stable is based on the result of Theorem 1.3.7 (Mañé's theorem) below. Let us start with a definition.

Fix a point $x \in M$ and consider the following two subspaces of T_xM:

$$B^+(x) = \left\{ v \in T_xM : \lim\inf_{k\to\infty} \left|Df^k(x)v\right| = 0 \right\}$$

and

$$B^-(x) = \left\{ v \in T_xM : \lim\inf_{k\to-\infty} \left|Df^k(x)v\right| = 0 \right\}.$$

Definition 1.3.11 We say that a diffeomorphism f satisfies the *analytic strong transversality condition* if

$$B^+(x) + B^-(x) = T_xM \quad \text{for any} \quad x \in M. \tag{1.18}$$

Theorem 1.3.7 *A diffeomorphism f is structurally stable if and only if f satisfies the analytic strong transversality condition.*

A detailed proof of Theorem 1.3.7 is given in Chap. 2 of this book.

Let us define one more important for us property of invariant sets of diffeomorphisms.

Let Λ be a compact invariant set of a diffeomorphism f.

Definition 1.3.12 We say that f admits a *dominated splitting on Λ* if there exist continuous families of linear subspaces $E(p)$ and $F(p)$ of the tangent spaces T_pM for $p \in \Lambda$ such that

(DS1) $E(p) \oplus F(p) = T_pM$, $p \in \Lambda$;

(DS2) the subspaces $E(p)$ and $F(p)$ are Df-invariant (i.e., analogs of equalities (HSD2.2) from Definition 1.3.1 with $S(p)$ and $U(p)$ replaced by $E(p)$ and $F(p)$ are satisfied);

(DS3) there exist numbers $C > 0$ and $\lambda \in (0, 1)$ such that

$$\left\|Df^k|_{E(p)}\right\| \cdot \left\|Df^{-k}|_{F(f^k(p))}\right\| \leq C\lambda^k, \quad p \in \Lambda, k \geq 0. \tag{1.19}$$

One more notion which we use in this book is the notion of a homoclinic point (and homoclinic trajectory).

Let p be a hyperbolic periodic point of a diffeomorphism f.

Definition 1.3.13 A point $q \neq p$ such that

$$q \in W^u(p) \cap W^s(p)$$

is called a *homoclinic* point of the periodic point p.

A homoclinic point q of p is called *transverse* if the stable and unstable manifolds $W^s(p)$ and $W^u(q)$ are transverse at q.

Theorem 1.3.8 *Any neighborhood of a transverse homoclinic point contains an infinite set of different hyperbolic periodic points of f.*

Many notions and statements formulated above for diffeomorphisms have analogs for flows generated by smooth vector fields. Let us give the corresponding definitions and state theorems which we need in what follows (in the case of similar objects, for example, such as the nonwandering set etc., we do not repeat the definitions and leave details to the reader).

Let X be a smooth (of class C^1) vector field on a smooth closed manifold M. Let

$$\phi : \mathbb{R} \times M \to M$$

be the flow generated by X and let, as above,

$$O(x, \phi) = \{\phi(t, x) : t \in \mathbb{R}\}$$

be the trajectory of a point $x \in M$ in the flow ϕ.

We denote by $\Phi(t, p)$ the derivative (in p) of $\phi(t, p)$; thus,

$$\Phi(t, p) : T_pM \to T_{\phi(t,p)}M.$$

Definition 1.3.14 We say that a set $I \subset M$ is a *hyperbolic set* of the vector field X (and its flow ϕ) if I has the following properties:

(HSF1) the set I is compact and ϕ-invariant;

(HSF2) there exist numbers $C > 0$ and $\lambda > 0$ and linear subspaces $S(p)$ and $U(p)$ of the tangent space T_pM defined for any point $p \in I$ such that

(HSF2.1) $S(p) \oplus U(p) \oplus \{X(p)\} = T_pM$, where $\{X(p)\}$ is the subspace spanned by the vector $X(p)$;

(HSF2.2)

$$\Phi(t,p)S(p) = S(\phi(t,p)) \quad \text{and} \quad \Phi(t,p)U(p) = U(\phi(t,p)), \quad t \in \mathbb{R};$$

(HSF2.3) if $v \in S(p)$, then $|\Phi(t,p)v| \le C\exp(-\lambda t)|v|$ for $t \ge 0$;
(HSF2.4) if $v \in U(p)$, then $|\Phi(t,p)v| \le C\exp(\lambda t)|v|$ for $t \le 0$.

Similarly to the case of diffeomorphisms, the main objects related to a hyperbolic set I of a flow ϕ are stable and unstable manifolds of its points (and its trajectories).

Definition 1.3.15 The *stable and unstable manifolds* of a point p are the sets defined by the equalities

$$W^s(p) = \{x \in M : \mathrm{dist}(\phi(t,x), \phi(t,p)) \to 0, \ t \to \infty\}$$

and

$$W^u(p) = \{x \in M : \mathrm{dist}(\phi(t,x), \phi(t,p)) \to 0, \ t \to -\infty\},$$

respectively.

One uses these objects to define the stable and unstable manifolds of the trajectory of a point p:

$$W^s(O(p,\phi)) = \bigcup_{t \in \mathbb{R}} W^s(\phi(t,p))$$

and

$$W^u(O(p,\phi)) = \bigcup_{t \in \mathbb{R}} W^u(\phi(t,p)).$$

The stable manifold theorem for flows states that if p is a point of a hyperbolic set I as above and $\sigma(p) = \dim S(p)$, then the structure of $W^s(O(p,\phi))$ can be described as follows:

- if p is a rest point (i.e., $\phi(t,p) \equiv p, \ t \in \mathbb{R}$), then $W^s(O(p,\phi)) = W^s(p)$ is the image of the Euclidean space $\mathbb{R}^{\sigma(p)}$ under a C^1 immersion;
- if $O(p,\phi)$ is a closed trajectory that is not a rest point (i.e., $\phi(t,p)$ is periodic in t with a nonzero minimal period), then $W^s(O(p,\phi))$ is the image under a C^1 immersion of a fiber bundle over the circle with fibers $\mathbb{R}^{\sigma(p)}$;
- if $O(p,\phi)$ is a trajectory such that $\phi(t_1,p) \ne \phi(t_2,p)$ for $t_1 \ne t_2$, then $W^s(O(p,\phi))$ is the image of the Euclidean space $\mathbb{R}^{\sigma(p)+1}$ under a C^1 immersion.

Similar statements hold for the unstable manifolds of trajectories of a hyperbolic set.

Now we recall the two basic definitions of the theory of structural stability of vector fields, the definitions of Ω-stability and structural stability.

Let us start with the definition of the C^1 topology on the space of vector fields of class C^1 on a smooth closed manifold M (everywhere below, a vector field is a vector field of class C^1).

Let X and Y be two such vector fields; define the number

$$\rho_1(X, Y) = \max_{x \in M} \left(|X(x) - Y(x)| + \left\| \frac{\partial X}{\partial x}(x) - \frac{\partial Y}{\partial x}(x) \right\| \right).$$

It is easily seen that ρ_1 is a metric on the space of vector fields of class C^1; we denote by $\mathscr{X}^1(M)$ (or simply by $\mathscr{X}^1$) the space of vector fields with this metric (and with the induced topology which we call C^1 topology). As in the case of diffeomorphisms, if A is a subset of $\mathscr{X}^1(M)$, then $\operatorname{Int}^1(A)$ denotes the interior of A in $\mathscr{X}^1(M)$.

Remark 1.3.7 Let X and Y be two vector fields and let ϕ and ψ be their flows, respectively. Consider the diffeomorphisms $f(x) = \phi(1, x)$ and $g(x) = \psi(1, x)$. It is not difficult to show that if $\rho_1(X, Y) \to 0$, then $\rho_1(f, g) \to 0$ (see, for example, Chap. 2 of [71]).

Let us denote by $\operatorname{Per}(X)$ (or $\operatorname{Per}(\phi)$) the set of rest points and closed trajectories of X (and its flow ϕ) and by $\Omega(X)$ ($\Omega(\phi)$) the nonwandering set (the definition of the nonwandering set for a flow is similar to that for a diffeomorphism, and we omit it).

Definition 1.3.16 A vector field X (and its flow ϕ) is called *structurally stable* if there exists a neighborhood W of X in $\mathscr{X}^1(M)$ such that for any vector field $Y \in W$, its flow ψ is *topologically equivalent* to the flow ϕ, i.e., there exists a homeomorphism $h : M \to M$ that maps trajectories of X to trajectories of Y preserving the orientation of trajectories.

Let us denote by $\mathscr{S}_F(M)$ (or $\mathscr{S}_F$) the set of structurally stable vector fields (and flows).

Remark 1.3.8 Let us note that, in contrast to Definition 1.3.4, it is not assumed in Definition 1.3.16 that h is a topological conjugacy of the flows ϕ and ψ of X and Y (the latter means that

$$h(\phi(t, x)) = \psi(t, h(x))$$

for all t and x).

In fact, the homeomorphism h in Definition 1.3.16 must have the following property: There exists a function $\tau : \mathbb{R} \times M \to \mathbb{R}$ such that

(1) for any $x \in M$, the function $\tau(\cdot, x)$ increases and maps $\mathbb{R}$ onto $\mathbb{R}$;
(2) $\tau(0, x) = x$ for any $x \in M$;
(3) $h(\phi(t, x)) = \psi(\tau(t, x), h(x))$ for any $(t, x) \in \mathbb{R} \times M$.

Clearly, the necessity of time reparametrization of shadowing trajectories in the case of shadowing for flows (see Sect. 1.2) is caused by the same reasons as the replacement of topological conjugacy by topological equivalence in Definition 1.3.16.

Definition 1.3.17 A vector field X (and its flow ϕ) is called *Ω-stable* if there exists a neighborhood W of X in $\mathscr{X}^1(M)$ such that for any vector field $Y \in W$, its flow ψ is *Ω-equivalent* to the flow ϕ, i.e., there exists a homeomorphism $h : \Omega(\phi) \to \Omega(\psi)$ that maps trajectories of $\Omega(\phi)$ to trajectories of $\Omega(\psi)$ preserving the orientation of trajectories.

Let us denote by $\Omega\mathscr{S}_F(M)$ (or $\Omega\mathscr{S}_F$) the set of Ω-stable vector fields (and flows).

The following condition (also introduced by Smale) is an analog of Axiom A for the case of vector fields and flows.

Axiom A′

(AA′a) The nonwandering set $\Omega(\phi)$ of the flow ϕ is hyperbolic.
(AA′b) The set $\Omega(\phi)$ is the union of two disjoint compact ϕ-invariant sets Q_1 and Q_2, where Q_1 consists of a finite number of rest points, while Q_2 does not contain rest points, and points of closed trajectories are dense in Q_2.

If a flow ϕ satisfies Axiom A′, then the following analog of Theorem 1.3.1 holds.

Theorem 1.3.9 *The nonwandering set $\Omega(\phi)$ has a unique representation of the form*

$$\Omega(\phi) = \Omega_1 \cup \cdots \cup \Omega_m,$$

where the Ω_i are disjoint, compact, ϕ-invariant sets such that each of these sets contains a dense positive semitrajectory.

As in the case of a diffeomorphism, the sets Ω_i are called *basic*. A basic set of a flow ϕ that satisfies Axiom A′ is either a rest point or a closed invariant set that does not contain rest points and such that points of closed trajectories are dense in it.

Let Ω_i and Ω_j be two different basic sets of a flow ϕ that satisfies Axiom A′. We write $\Omega_i \to \Omega_j$ if there exists a point x such that

$$\phi(t,x) \to \Omega_i,\ t \to -\infty, \quad \text{and} \quad \phi(t,x) \to \Omega_j,\ t \to \infty.$$

The no cycle condition for a flow ϕ literally repeats the corresponding condition for a diffeomorphism.

The following statement is an analog of Theorem 1.3.3.

Theorem 1.3.10 *A flow ϕ is Ω-stable if and only if ϕ satisfies Axiom A′ and the no cycle condition.*

If a flow ϕ satisfies Axiom A′, then hyperbolic trajectories $\phi(t,p)$, $p \in \Omega(\phi)$, have stable and unstable manifolds $W^s(O(p,\phi))$ and $W^u(O(p,\phi))$, respectively.

Definition 1.3.18 We say that such a flow ϕ satisfies the *geometric strong transversality condition* if for any points $p, q \in \Omega(\phi)$, the manifolds $W^s(O(q,\phi))$ and $W^u(O(p,\phi))$ are transverse at any point of their intersection.

The following statement is an analog of Theorem 1.3.4.

Theorem 1.3.11 *A flow ϕ is structurally stable if and only if ϕ satisfies Axiom A′ and the geometric strong transversality condition.*

Definition 1.3.19 A rest point or a closed trajectory of a flow ϕ is called *hyperbolic* if it is a hyperbolic set of ϕ.

Remark 1.3.9 Condition under which a rest point or a closed trajectory is hyperbolic are well-known:

- a rest point p of a flow ϕ generated by a vector field X is hyperbolic if and only if any eigenvalue of the Jacobi matrix $DX(p)$ has nonzero real part;
- a closed trajectory γ of a flow ϕ is hyperbolic if and only if, for any transverse section Σ at any point of γ, the zero point of the section (corresponding to the intersection of γ with Σ) is a hyperbolic fixed point of the Poincaré map generated by Σ (see [71] for details).

Denote by HP_F the set of flows ϕ such that any rest point and closed trajectory of ϕ is hyperbolic.

A complete analog of Theorem 1.3.5 for vector fields (and flows) is not correct (see Historical Remarks at the end of this section). Only the following partial analog is valid.

Theorem 1.3.12 *A nonsingular vector field in $Int^1(HP_F)$ belongs to $\Omega\mathscr{S}_F$.*

Definition 1.3.20 A flow $\phi \in \mathrm{HP}_F$ is called *Kupka–Smale* if stable and unstable manifolds of its rest points and closed trajectories are transverse. We denote by KS_F the set of Kupka–Smale flows.

The following statement is an analog of Theorem 1.3.6.

Theorem 1.3.13

(1) The set KS_F is residual in $\mathscr{X}^1$.
(2) The sets $Int^1(KS_F)$ and $\mathscr{S}_F$ coincide.

Let us describe one more approach for establishing the structural stability of a flow.

Let, as above, ϕ be the flow generated by a vector field X.

Definition 1.3.21 A point $x \in M$ is called a *chain recurrent point* of the flow ϕ if for any $d, T > 0$ there exists a d-pseudotrajectory g of ϕ (in the sense of Definition 1.2.1) such that $g(0) = x$ and $g(t) = x$ for some $t \geq T$.

In this case, similarly to Sect. 1.1, we write $x \leftrightsquigarrow x$.

Definition 1.3.22 The set

$$\mathscr{R}(\phi) = \{x \in M : x \leftrightsquigarrow x\}$$

of all chain recurrent points of ϕ is called the *chain recurrent set of* ϕ.

It is easy to show (compare with Sect. 1.1) that in our case (where M is a compact manifold), the set $\mathscr{R}(\phi)$ is nonempty, compact, and ϕ-invariant.

In Sect. 2.7, we refer to the following two results.

Theorem 1.3.14 *If X is a vector field of class C^1 such that the chain recurrent set $\mathscr{R}(\phi)$ of its flow ϕ is hyperbolic and stable and unstable manifolds of trajectories in $\mathscr{R}(\phi)$ are transverse, then X is structurally stable.*

Now we formulate a theorem which allows one to show that components of the set $\mathscr{R}(\phi)$ are hyperbolic.

Let Σ be a compact, ϕ-invariant component of $\mathscr{R}(\phi)$ that does not contain rest points of ϕ. Denote $f(x) = \phi(1, x)$.

For a point $x \in \Sigma$, denote by $P(x)$ the orthogonal projection in T_xM with kernel spanned by $X(x)$ and by $V(x)$ the orthogonal complement to $X(x)$ in T_xM. Consider the normal subbundle $\mathscr{V}(\Sigma)$ of the tangent bundle $TM|_\Sigma$ which is the set of pairs $(x, V(x))$, where $x \in \Sigma$.

Define a mapping π on the normal subbundle $\mathscr{V}(\Sigma)$ over Σ by

$$\pi(x, v) = (f(x), B(x)v)\,, \text{ where } B(x) = P(f(x))Df(x)$$

(recall that $f(x) = \phi(1, x)$).

The hyperbolicity of π on $\mathscr{V}(\Sigma)$ is defined similarly to the usual hyperbolicity of a diffeomorphism on a compact invariant set. It means that there exist numbers $C > 0$ and $\lambda \in (0, 1)$ and linear subspaces $S(p), U(p)$ of $V(p)$ for $p \in \Sigma$ such that

- $S(p) \oplus U(p) = V(p)$;
- $B(p)S(p) = S(f(p))$ and $B(p)U(p) = U(f(p))$;
- if $v \in S(p)$, then $|B^k(p)v| \le C\lambda^k|v|$ for $k \ge 0$;
- if $v \in U(p)$, then $|B^k(p)v| \le C\lambda^{-k}|v|$ for $k \le 0$.

Theorem 1.3.15 *If π is hyperbolic on $\mathscr{V}(\Sigma)$, then Σ is a hyperbolic set of the flow ϕ.*

If p is a rest point of a flow ϕ (i.e., $O(p, \phi) = \{p\}$), then we denote by $W^s(p)$ and $W^u(p)$ (instead of $W^s(O(p, \phi))$ etc.) its stable and unstable manifolds, respectively.

If γ is a closed trajectory of a flow ϕ (i.e., $O(p, \phi) = \gamma$ for any $p \in \gamma$), then we denote by $W^s(\gamma)$ and $W^u(\gamma)$ its stable and unstable manifolds, respectively.

Let p be a hyperbolic rest point (or let γ be a hyperbolic closed trajectory) of a flow ϕ.

Definition 1.3.23 A point $q \neq p$ such that

$$q \in W^u(p) \cap W^s(p)$$

is called a *homoclinic* point of the rest point p.

A point $q \notin \gamma$ such that

$$q \in W^u(\gamma) \cap W^s(\gamma)$$

is called a *homoclinic* point of the closed trajectory γ.

A homoclinic point q of γ is called *transverse* if the stable and unstable manifolds $W^s(p)$ and $W^u(q)$ are transverse at q.

Remark 1.3.10 Let us note that a homoclinic point q of a hyperbolic rest point p cannot be transverse. Indeed, such a point q cannot be a rest point (otherwise, $q = p$); hence, $X(q) \neq 0$ (where X is the vector field which generates the flow ϕ).

Since

$$\dim W^s(p) + \dim W^u(p) = \dim M$$

and

$$0 \neq X(q) \in T_q W^s(p) \cap T_q W^u(p),$$

the equality

$$T_q W^s(p) + T_q W^s(p) = T_q M$$

is impossible.

An analog of Theorem 1.3.8 for flows can be formulated as follows.

Theorem 1.3.16 *If q is a transverse homoclinic point of a hyperbolic closed trajectory γ of a flow ϕ, then any neighborhood of $O(q, \phi)$ contains an infinite set of different hyperbolic closed trajectories of ϕ.*

Historical Remarks The general definition of a hyperbolic set is usually attributed to D. V. Anosov [3].

The stable manifold theorem has a long history; usually, one refers to the names of J. Hadamard and O. Perron (one can find an interesting discussion concerning the theory of stable and unstable manifolds in D. V. Anosov's monograph [3]; there he mentiones also G. Darboux, H. Poincaré, and A. M. Lyapunov).

The notions of nonwandering points and other classical objects of the global theory of dynamical systems were introduced and studied by G. Birkhoff [10].

The theory of structural stability originates from the A. A. Andronov and L. S. Pontryagin's paper [2] in which they defined a kind of such a property for vector fields in a two-dimensional disk or on the two-dimensional sphere.

A very important role was played by S. Smale's paper [95] in which he introduced the notions of Ω-stability, Axioms A and A′, proved the spectral decomposition theorem (Theorem 1.3.1), gave first sufficient conditions of Ω-stability, etc.

Later, S. Smale proved the sufficiency of conditions of Theorem 1.3.3 [98].

The basic results of the theory of Ω-stability and structural stability were formulated as conjectures by J. Palis and S. Smale [52].

The sufficiency statement in Theorem 1.3.4 was first proved by J. Robbin in [78] for diffeomorphisms of class C^2 and later by C. Robinson [81] in the general case.

The necessity of conditions of Theorem 1.3.4 was established by R. Mañé in [45]; later, the necessity of conditions of Theorem 1.3.5 was proved by J. Palis [53].

The set HP was studied by many authors; the set $\mathrm{Int}^1(\mathrm{HP})$ is sometimes denoted by $\mathscr{F}$ (or $\mathscr{F}^1$), and its elements are called star systems (both in the case of diffeomorphisms and in the case of vector fields).

Theorem 1.3.5 was proved by Aoki [7] and S. Hayashi [25].

The complete analog of Theorem 1.3.5 for vector fields (and flows) is not correct. A vector field in $\mathrm{Int}^1(\mathrm{HP}_F)$ may fail to have a hyperbolic nonwandering set, as the famous Lorenz attractor shows [22], or fail to have rest points and closed trajectories dense in the nonwandering set [17], or, even if Axiom A′ is satisfied, still fail to satisfy the no cycle condition [37].

R. Mañé proved Theorem 1.3.7 in [39].

Theorem 1.3.12 was proved S. Gan and L. Wen in [21].

Kupka–Smale systems were independently defined and studied by I. Kupka [31] and S. Smale [94]. They proved Theorem 1.3.6 (1) and Theorem 1.3.13 (1).

Theorem 1.3.6 (2) follows from the results of [7] (where it was proved that $\mathrm{Int}^1(\mathrm{KS}_D) \subset \mathscr{S}_D$) and [82], where the inverse inclusion was established.

The inclusion $\mathrm{Int}^1(\mathrm{KS}_F) \subset \mathscr{S}_F$ was proved by H. Toyoshiba [103] and C. Robinson [80]; the inverse inclusion was established by C. Robinson [79] and S. Hayashi [26].

Homoclinic points were first studied by H. Poincaré [75]; Theorem 1.3.8 (as well as Theorem 1.3.13) belongs to S. Smale [96, 97].

The sufficiency of conditions of Theorem 1.3.10 was established by C. Pugh and M. Shub [76]; the sufficiency of conditions of Theorem 1.3.11 was proved by C. Robinson [79].

The necessity of conditions in these theorems follows from results of L. Wen [106] and S. Hayashi [26].

It was shown by J. E. Franke and J. F. Selgrade in [18] that for a flow ϕ, the set $\mathscr{R}(\phi)$ is hyperbolic if and only if ϕ satisfies Axiom A′ and the no cycle condition. Theorem 1.3.14 follows from this result combined with Theorem 1.3.11.

R. Sacker and G. Sell studied in detail dichotomies and invariant splittings in linear differential systems [86]; in particular, they proved Theorem 1.3.15 in [85].

1.4 Hyperbolic Shadowing

As we wrote in the Preface, one of the main goals of this book is to study relations between shadowing and basic notions of the theory of structural stability. It was known that structural stability implies Lipschitz shadowing both for diffeomorphisms and vector fields; let us formulate this as a theorem.

Theorem 1.4.1 *The following inclusions hold:*

(1) $\mathscr{S}_D \subset LSP_D$*;*
(2) $\mathscr{S}_F \subset LSP_F$*.*

We show in Chap. 2 that the inverse inclusions hold as well, so that structural stability is equivalent to Lipschitz shadowing.

An important part in the proof of Theorem 1.4.1 is the statement that a diffeomorphism or a vector field has the Lipschitz shadowing property in a neighborhood of its hyperbolic set.

In this section, we prove that a diffeomorphism has the finite Lipschitz shadowing property in a neighborhood of a hyperbolic set (in this book, we refer to this statement in Sect. 2.4). This is a classical result having a lot of different proofs. The proof which we give here is of a geometric origin; its modification can be applied in the absence of hyperbolicity as well (see, for example, [58]).

To simplify presentation, we consider a diffeomorphism f of $\mathbb{R}^n$ and its hyperbolic set Λ.

Our proof applies the existence of a so-called *adapted* (or *Lyapunov*) norm in a neighborhood of Λ (with respect to this norm, the constant C in inequalities (HSD2.3) and (HSD2.4) of Sect. 1.3 equals 1); a proof of this result can be found in [71].

Lemma 1.4.1 *Let Λ be a hyperbolic set of a diffeomorphism f. There exist constants $\nu \geq 1$ and $\lambda \in (0, 1)$ such that for any $\varepsilon > 0$ we can find a neighborhood W of the set Λ having the following property. There exists a positive constant δ, a C^∞ norm $|\cdot|_x$ for $x \in W$, and continuous (but not necessarily Df-invariant) extensions S' and U' of the families S and U of the given hyperbolic structure to the neighborhood W such that*

(1) $S'(p) \oplus U'(p) = \mathbb{R}^n, \quad p \in W;$

(2) if $p, q \in W$, $|f(p) - q| \leq \delta$, and $P(q)$ is the projection onto $S'(q)$ parallel to $U'(q)$, then the mapping $P(q)Df(p)$ is a linear isomorphism between $S'(p)$ and $S'(q)$ (respectively, if $Q(q) = Id - P(q)$, then the mapping $Q(q)Df(p)$ is a linear isomorphism between $U'(p)$ and $U'(q)$) and the following inequalities hold:

$$|P(q)Df(p)v|_q \leq \lambda |v|_p \text{ and } |Q(q)Df(p)v|_q \leq \varepsilon |v|_p, \quad v \in S'(p), \tag{1.20}$$

and

$$\lambda |Q(q)Df(p)v|_q \geq |v|_p \text{ and } |P(q)Df(p)v|_q \leq \varepsilon |v|_p, \quad v \in U'(p); \tag{1.21}$$

(3)

$$\frac{1}{\nu}|v|_p \leq |v| \leq \nu |v|_p, \quad p \in W, \ v \in \mathbb{R}^n; \tag{1.22}$$

(4)

$$\| P(p)\|, \|Q(p)\| \leq \nu, \quad p \in W \tag{1.23}$$

(in inequalities (1.23), we have in mind the operator norm related to the norm $|.|_p$).

Remark 1.4.1

(1) Since the adapted norm is Lipschitz equivalent to the standard norm (see inequalities (1.22)), f has (or does not have) the finite Lipschitz shadowing property with respect to these norms simultaneously. For that reason, to simplify presentation, we assume that the standard Euclidean norm is adapted. Similarly, we write $S(p)$ and $Q(p)$ instead of $S'(p)$ and $U'(p)$ for $p \in W$.
(2) In addition, we may assume that the neighborhoods W corresponding to small enough ε are subsets of a fixed closed neighborhood of Λ.

This allows us to assume that the norm $\|Df(p)\|$ is bounded for $p \in W$ and to use uniform estimates of the remainder term of the first-order Taylor formula for f in the proof of property (P′4) and in formula (1.35).

Thus, we assume that

$$\|Df(p)\| \le M_0, \quad p \in W,$$

and set $M = \nu(1 + 12M_0)$.

Take

$$\mathcal{L} = 2\nu/(1 - \lambda) \tag{1.24}$$

and note that

$$\mathcal{L} > \lambda\mathcal{L} + \nu > 1 \text{ and } \mathcal{L}/\lambda > \mathcal{L} + \nu. \tag{1.25}$$

There exists an $\varepsilon > 0$ such that

$$\mathcal{L} > \nu + \varepsilon M(1 + \nu)\mathcal{L}, \tag{1.26}$$

$$\mathcal{L} > \lambda\mathcal{L} + \nu + \varepsilon(1 + 2\nu)\mathcal{L}. \tag{1.27}$$

Note that (1.27) implies the inequality

$$\mathcal{L}/\lambda > \mathcal{L} + \nu + \varepsilon(1 + 2\nu)\mathcal{L}. \tag{1.28}$$

Let W be a neighborhood of Λ corresponding by Lemma 1.4.1 to this ε.

Our main result in this section is as follows.

Theorem 1.4.2 *The diffeomorphism f has the finite Lipschitz shadowing property in W.*

Proof First we define several geometric objects related to the introduced structure.

Fix a point p in W; we represent points q close to p in the form $p + v$ and define our objects by imposing conditions on the projections $P(p)v$ and $Q(p)v$.

Let Δ' and Δ be positive numbers; consider the sets

$$R(\Delta', \Delta, p) = \{q = p + v : |P(p)v| \le \Delta', |Q(p)v| \le \Delta\};$$

we write $R(\Delta, p)$ instead of $R(\Delta, \Delta, p)$. Let

$$V(\Delta, p) = \{q = p + v \in R(\Delta, p) : |Q(p)v| = \Delta\}$$

and

$$T(\Delta, p) = \{q = p + v \in R(\Delta, p) : Q(p)v = 0\}.$$

Let us note several obvious geometric properties of the introduced objects.

(P1) $V(\Delta, p)$ is not a retract of $R(\Delta, p)$.
(P2) $V(\Delta, p)$ is a retract of $R(\Delta, p) \setminus T(\Delta, p)$.
(P3) If $\Delta' > \Delta$, then there exists a retraction

$$\sigma : R(\Delta', p) \to R(\Delta, p)$$

such that if

$$q = p + v \text{ and } Q(p)v \neq 0,$$

then

$$\sigma(q) = p + v', \text{ where } Q(p)v' \neq 0.$$

Now we prove several properties of the images of the introduced sets under f.

(P4) There exists a $\Delta_1 > 0$ such that if $p, r, f(p) \in W$, $\Delta \le \Delta_1$, and $|r - f(p)| < \Delta$, then

$$f(R(\Delta, p)) \subset R(M_1\Delta, r) \tag{1.29}$$

and

$$f^{-1}(R(\Delta, r)) \subset R(M_1\Delta, p), \tag{1.30}$$

where $M_1 = 4\nu M_0$.

We prove only the part of property (P4) related to inclusion (1.29); the part related to inclusion (1.30) is proved by a similar reasoning (possibly, with different constants M_1 and Δ_1).

First we prove an auxiliary statement:

(P4$'$) There exists a $\Delta_1 > 0$ such that if $p, f(p) \in W$ and $\Delta \le \Delta_1$, then

$$f(R(\Delta, p)) \subset R(M_1\Delta, f(p)), \tag{1.31}$$

where $M_1 = 4\nu M_0$.

Indeed, take a point $q = p + v \in R(\Delta, p)$; then

$$|v| \le |P(p)v| + |Q(p)v| \le 2\Delta.$$

Since

$$f(q) = f(p) + Df(p)v + o(p, v),$$

where

$$|o(p, v)|/|v| \to 0, \quad |v| \to 0,$$

uniformly in p and $\|Df(p)\| \le M_0$, there exists a $\Delta_1 > 0$ such that if $\Delta \le \Delta_1$, then

$$|f(q) - f(p)| \le 2M_0|v|, \quad |v| \le 2\Delta.$$

If $f(q) = f(p) + w$, then

$$|P(f(p))w|, |Q(f(p))w| \le 2\nu M_0|v| \le 4\nu M_0\Delta,$$

which proves (P4$'$) with $M_1 = 4\nu M_0$.

Now we prove (1.29). Since the projections P and Q are uniformly continuous, we can reduce, if necessary, Δ_1 so that

$$\|P(x) - P(y)\|, \|Q(x) - Q(y)\| < 1, \quad x, y \in W, \ |x - y| < \Delta_1. \tag{1.32}$$

Let $\Delta \le \Delta_1$. Take a point $q \in f(R(\Delta, p))$ and let

$$q = f(p) + v = r + w.$$

Then $|v - w| < \Delta$ and

$$|P(f(p))v|, |Q(f(p))v| \le M_1\Delta$$

by (P4$'$).

Let us estimate

$$|P(r)w| \le |P(r)w - P(r)v| + |P(r)v - P(f(p))v| + |P(f(p))v| \le$$

$$\le \nu\Delta + 2M_1\Delta + M_1\Delta = (\nu + 3M_1)\Delta = M\Delta$$

(estimating the second term, we take the inequality $|v| \leq 2M_1\Delta$ and (1.32) into account).

A similar estimate holds for $|Q(r)w|$, which proves (1.29).

Of course, without loss of generality, we may assume that

$$M \geq 1. \tag{1.33}$$

Now we fix a

$$d_0 \in (0, \Delta_1/\mathcal{L})$$

with the following properties:

(1) if $p, f(p), r \in W$ and $|r - f(p)| < d_0$, then inequalities (1.20) and (1.21) are satisfied with the chosen ε;

(2) in the representation

$$f(p+v) = f(p) + Df(p)v + o(p,v), \tag{1.34}$$

the estimate

$$|o(p,v)| \leq \varepsilon|v|, \quad |v| \leq 2M\mathcal{L}d_0, \tag{1.35}$$

holds.

Now we prove one more statement.

(P5) If $d \leq d_0$, $p, f(p), r \in W$, $|r - f(p)| < d$, and $\Delta = \mathcal{L}d$, then

$$f(T(M\Delta, p)) \cap V(\Delta, r) = \emptyset, \tag{1.36}$$

$$f(T(\Delta, p)) \subset \mathrm{Int}(R(\Delta, r)), \tag{1.37}$$

$$f(R(\Delta, p)) \cap \partial R(\Delta, r) \subset V(\Delta, r), \tag{1.38}$$

and

$$f(V(\Delta, p)) \cap R(\Delta, r) = \emptyset. \tag{1.39}$$

First we prove relation (1.36).

If $q = p + v \in T(M\Delta, p)$, then $v = P(p)v \in S(p)$, $|P(p)v| \leq M\Delta = M\mathcal{L}d$, and $Q(p)v = 0$. Hence, it follows from representation (1.34) and estimates (1.26) and (1.35) that

$$|Q(r)(f(q) - r)| \leq |Q(r)(f(p) - r)| + |Q(r)Df(p)P(p)v| + |Q(r)o(p,v)| \leq$$

$$\leq \nu d + \varepsilon M\mathcal{L}d + \nu\varepsilon M\mathcal{L}d = (\nu + \varepsilon M(1+\nu)\mathcal{L})d < \mathcal{L}d = \Delta,$$

which proves relation (1.36).

Let us prove relations (1.37) and (1.38).

First we note that inequality (1.33) implies the inclusion

$$T(\Delta, p) \subset T(M\Delta, p),$$

and it follows from the above inequality that

$$|Q(r)(f(q) - r)| < \Delta, \quad q \in T(\Delta, p). \tag{1.40}$$

Now we consider a point $q = p + v \in R(\Delta, p)$, represent $v = P(p)v + Q(p)v$, and estimate

$$|P(r)(f(q) - r)| \leq |P(r)(f(p) - r)| + |P(r)Df(p)P(p)v| +$$

$$+|P(r)Df(p)Q(p)v| + |P(r)o(p, v)| \leq$$

$$\leq \nu d + \lambda \mathscr{L} d + \varepsilon \mathscr{L} d + 2\nu\varepsilon \mathscr{L} d = (\nu + \lambda \mathscr{L} + \varepsilon(1 + 2\nu)\mathscr{L})d < \mathscr{L} d = \Delta$$

(here we refer to the estimate $|v| \leq 2\mathscr{L} d$ and to inequality (1.27)).

The above inequality proves relation (1.38). Combining it with inequality (1.40), we get a proof of relation (1.37).

Finally, we prove relation (1.39). If $q = p + v \in V(\Delta, p)$, then $|P(p)v| \leq \Delta = \mathscr{L} d$ and $|Q(p)v| = \Delta = \mathscr{L} d$. Then

$$|Q(r)(f(q) - r)| \geq$$

$$\geq |Q(r)Df(p)(P(p)v + Q(p)v)| - |Q(r)(f(p) - r)| - |Q(r)o(p, v)| \geq$$

$$\geq |Q(r)Df(p)Q(p)v| - |Q(r)Df(p)Q(p)v| - |Q(r)(f(p) - r)| - |Q(r)o(p, v)| \geq$$

$$\geq \mathscr{L} d/\lambda - \varepsilon \mathscr{L} d - \nu d - 2\varepsilon\nu \mathscr{L} d = (\mathscr{L}/\lambda - \nu - \varepsilon(1 + 2\nu))d > \mathscr{L} d = \Delta$$

(here we refer to inequality (1.28)). This proves relation (1.39).

Now we consider points $p_0, \dots, p_m \in W$ such that

$$f(p_k) \in W, \quad k = 0, \dots, m - 1, \tag{1.41}$$

and

$$|f(p_k) - p_{k+1}| < d \leq d_0, \quad k = 0, \dots, m - 1,$$

and prove that there exists a point $r \in R(\Delta, p_0)$ such that

$$f^k(r) \in R(\Delta, p_k), \quad k = 1, \dots, m, \tag{1.42}$$

where $\Delta = \mathscr{L}d$.

Let us note that condition (1.41) is not a real restriction since we can guarantee it reducing W, if necessary.

For brevity, we denote $R_k = R(\Delta, p_k), V_k = V(\Delta, p_k), T_k = T(\Delta, p_k)$.

Consider the sets

$$A_k = R_k \setminus \bigcap_{l=k+1}^{m} f^{-(l-k)}(\mathrm{Int}(R_l)), \quad k = 0, \dots, m-1.$$

It follows from equality (1.39) that

$$f(V_k) \cap R_{k+1} = \emptyset.$$

Hence, $V_k \subset A_k$.

We claim that there exist retractions

$$\rho_k : A_k \to V_k, \quad k = 0, \dots, m-1.$$

This is enough to prove our statement since the existence of ρ_0 means that

$$\bigcap_{l=0}^{m} f^{-l}(\mathrm{Int}(R_l)) \neq \emptyset$$

(otherwise there exists a retraction of R_0 to V_0, which is impossible by property (P1)), which, in turn, means that there exists a point $r \in R_0$ such that

$$f^k(r) \in R_k, \quad k = 0, \dots, m,$$

or

$$\left| f^k(r) - p_k \right| \leq 2\nu\mathscr{L}d, \quad k = 0, \dots, m.$$

Thus, our claim implies the finite Lipschitz shadowing property of f in W with constants d_0 and $2\nu\mathscr{L}$.

Let us prove our claim. The existence of ρ_{m-1} is obvious since inclusion (1.37) implies that

$$T_{m-1} \subset f^{-1}(\mathrm{Int}(R_m)),$$

and hence,

$$R_{m-1} \setminus f^{-1}(\mathrm{Int}(R_m)) \subset R_{m-1} \setminus T_{m-1},$$

while V_{m-1} is a retract of the latter set by property (P2).

Let us assume that the existence of retractions $\rho_{k+1}, \ldots, \rho_{m-1}$ has been proved. Let us prove the existence of ρ_k.

The definition of the sets A_k implies that

$$A_k \cap f^{-1}(R_{k+1}) \subset f^{-1}(A_{k+1}) \tag{1.43}$$

since

$$f(A_k) \cap f^{-(l-k)+1}((\mathrm{Int}(R_l)) = \emptyset \text{ for } l \geq k+2.$$

Define a mapping θ on A_k by setting

$$\theta(q) = f^{-1} \circ \rho_{k+1} \circ f(q), \quad q \in A_k \cap f^{-1}(R_{k+1}),$$

and

$$\theta(q) = q, \quad q \in A_k \setminus f^{-1}(R_{k+1}).$$

Inclusion (1.43) shows that the mapping θ is properly defined.

Let us show that this mapping is continuous. Clearly, it is enough to show that $\rho_{k+1}(r) = r$ for $r \in f(A_k \cap f^{-1}(\partial R_{k+1}))$.

For this purpose, we note that

$$f(A_k \cap f^{-1}(\partial R_{k+1})) = f(A_k) \cap \partial R_{k+1} \subset f(R_k) \cap \partial R_{k+1} \subset V_{k+1}$$

(we refer to inclusion (1.38)) and $\rho_{k+1}(r) = r$ for $r \in V_{k+1}$.

Clearly, θ maps A_k into the set

$$B_k = [R_k \setminus f^{-1}(R_{k+1})] \cup f^{-1}(V_{k+1}). \tag{1.44}$$

Since $d < \Delta_1$ by our choice of d_0, it follows from property (P4) that

$$B_k \subset R(M\Delta, p_k).$$

Let us consider a retraction

$$\sigma : R(M\Delta, p_k) \to R_k$$

given by property (P3).

If

$$q = p_k + v \in f^{-1}(V_{k+1}) \setminus R_k,$$

then $q \notin T(M\Delta, p_k)$ by (1.36); thus, $Q(p)v \neq 0$. It follows from property (P3) that in this case,

$$\sigma(q) \in C_k := R_k \setminus T(\Delta, p_k).$$

If

$$q \in R_k \setminus f^{-1}(R_{k+1}),$$

then the above inclusion follows from (1.37).

Condition (P2) implies that there exists a retraction

$$\rho : \ C_k \to V_k.$$

It remains to note that $\theta(q) = q$ for $q \in V_k$ due to relation (1.39). Thus,

$$\rho_k = \rho \circ \sigma \circ \theta : \ A_k \to V_k$$

is the required retraction. □

Historical Remarks There exist several proofs of the inclusion

$$\mathscr{S}_D \subset \mathrm{SSP}_D$$

based on different ideas.

This statement was proved by A. Morimoto [46], K. Sawada [92], and C. Robinson [83] (note that the proof in [83] is not complete).

As far as the authors know, the first statement of Theorem 1.4.1 was first proved in the book [61] of the first author, and the second statement was proved in his paper [62].

Lemma 1.4.1 belongs to D. V. Anosov [3].

As was mentioned in Historical Remarks to Sect. 1.1, both classical proofs of the shadowing property in a neighborhood of a hyperbolic set of a diffeomorphism given by D. V. Anosov in [4] and R. Bowen in [12] show that shadowing is Lipschitz.

Our proof of Theorem 1.4.2 published in the joint paper [58] of the first author and A. A. Petrov mostly follows the ideas of the joint paper [63] of the first author and O. B. Plamenevskaya.

Chapter 2
Lipschitz and Hölder Shadowing and Structural Stability

In this chapter, we give either complete proofs or schemes of proof of the following main results:

- If a diffeomorphism f of a smooth closed manifold has the Lipschitz shadowing property, then f is structurally stable (Theorem 2.3.1);
- a diffeomorphism f has the Lipschitz periodic shadowing property if and only if f is Ω-stable (Theorem 2.4.1);
- if a diffeomorphism f of class C^2 has the Hölder shadowing property on finite intervals with constants $\mathscr{L}, C, d_0, \theta, \omega$, where $\theta \in (1/2, 1)$ and $\theta + \omega > 1$, then f is structurally stable (Theorem 2.5.1);
- there exists a homeomorphism of the interval that has the Lipschitz shadowing property and a nonisolated fixed point (Theorem 2.6.1);
- if a vector field X has the Lipschitz shadowing property, then X is structurally stable (Theorem 2.7.1).

The structure of the chapter is as follows.

We devote Sects. 2.1–2.3 to the proof of Theorem 2.3.1. In Sect. 2.1, we prove theorems of Maizel' and Pliss relating the so-called Perron property of difference equations and hyperbolicity of sequences of linear automorphisms, Sect. 2.2 is devoted to the Mañé theorem (Theorem 1.3.7), and in Sect. 2.3, we reduce the proof of Theorem 2.3.1 to results of the previous two sections.

Theorem 2.4.1 is proved in Sect. 2.4; Theorem 2.5.1 is proved in Sect. 2.5; Theorem 2.6.1 is proved in Sect. 2.6.

Finally, Sect. 2.7 is devoted to the proof of Theorem 2.7.1.

S.Yu. Pilyugin, K. Sakai, *Shadowing and Hyperbolicity*, Lecture Notes in Mathematics 2193, DOI 10.1007/978-3-319-65184-2_2

2.1 Maizel' and Pliss Theorems

Let $I = \{k \in \mathbb{Z} : \; k \geq 0\}$. Let $\mathscr{A} = \{A_k, \; k \in I\}$ be a sequence of linear isomorphisms

$$A_k : \mathbb{R}^n \to \mathbb{R}^n.$$

We assume that there exists a constant $N \geq 1$ such that

$$\|A_k\|, \|A_k^{-1}\| \leq N, \quad k \in I. \tag{2.1}$$

We relate to this sequence two difference equations, the homogeneous one,

$$x_{k+1} = A_k x_k, \quad k \in I, \tag{2.2}$$

and the inhomogeneous one,

$$x_{k+1} = A_k x_k + f_{k+1}, \quad k \in I. \tag{2.3}$$

Definition 2.1.1 We say that the sequence $\mathscr{A}$ has the *Perron property* on I if for any bounded sequence f_k, Eq. (2.3) has a bounded solution.

Set

$$F(k,l) = \begin{cases} A_{k-1} \circ \cdots \circ A_l, & k > l; \\ \text{Id}, & k = l; \\ A_k^{-1} \circ \cdots \circ A_{l-1}^{-1}, & k < l. \end{cases}$$

Definition 2.1.2 We say that the sequence $\mathscr{A}$ is *hyperbolic* on I if there exist constants $C > 0$ and $\lambda \in (0,1)$ and projections $P_k, Q_k, k \in I$, such that if $S_k = P_k\mathbb{R}^n$ and $U_k = Q_k\mathbb{R}^n$, then

$$S_k \oplus U_k = \mathbb{R}^n; \tag{2.4}$$

$$A_k S_k = S_{k+1}, \quad A_k U_k = U_{k+1}; \tag{2.5}$$

$$|F(k,l)v| \leq C\lambda^{k-l}|v|, \quad v \in S_l, \; k \geq l; \tag{2.6}$$

$$|F(k,l)v| \leq C\lambda^{l-k}|v|, \quad v \in U_l, \; k \leq l; \tag{2.7}$$

$$\|P_k\|, \|Q_k\| \leq C. \tag{2.8}$$

In the relations above, $k, l \in I$.

Our first main result in this section is the following statement.

Theorem 2.1.1 (Maizel') *If the sequence $\mathscr{A}$ has the Perron property on I, then this sequence is hyperbolic on I.*

Remark 2.1.1 Of course, it is well known that a hyperbolic sequence $\mathscr{A}$ has the Perron property on I (see Lemma 2.1.6 below), so the properties of $\mathscr{A}$ in the above theorem are equivalent. We formulate it in the above form since this implication is what we really need (and since precisely this statement was proved by Maizel').

Proof Thus, we assume that the sequence $\mathscr{A}$ has the Perron property on I.

Let us denote by $\mathscr{B}$ the Banach space of bounded sequences $x = \{x_k\}$, where $x_k \in \mathbb{R}^n$ and $k \in I$, with the usual norm

$$\|x\| = \sup_{k \in I} |x_k|.$$

A sequence $x \in \mathscr{B}$ that satisfies Eq. (2.2) (or (2.3)) will be called a $\mathscr{B}$-solution of the corresponding equation.

Denote

$$V_1 = \{x_0 : x = (x_0, x_1, \dots) \quad \text{is a } \mathscr{B} - \text{solution of} \quad (2.2)\}.$$

Since Eq. (2.2) is linear and $\mathscr{B}$ is a linear space, V_1 is a linear space as well. Denote by V_2 the orthogonal complement of V_1 in $\mathbb{R}^n$ and by P the orthogonal projection to V_1.

The difference of any two $\mathscr{B}$-solutions of Eq. (2.3) with a fixed $f \in \mathscr{B}$ is a $\mathscr{B}$-solution of Eq. (2.2). It is easily seen that for any $f \in \mathscr{B}$ there exists a unique $\mathscr{B}$-solution of Eq. (2.3) (we denote it $T(f)$) such that $(T(f))_0 \in V_2$.

The defined operator

$$T : \mathscr{B} \to \mathscr{B}$$

plays an important role in the proof. Clearly, the operator T is linear.

Lemma 2.1.1 *The operator T is continuous.*

Proof Since we know that the operator T is linear, it is enough to show that the graph of T is closed; then our statement follows from the closed graph theorem.

Thus, assume that

$$f_n = (f_0^n, \dots) \in \mathscr{B}, \quad y_n = (y_0^n, \dots) \in \mathscr{B},$$

$y_n = T(f_n), f_n \to f$, and $y_n \to y = (y_0, \dots)$ in $\mathscr{B}$.

Then, clearly, $y_0 \in V_2$.
Fix $k \in I$ and pass in the equality

$$y_{k+1}^n = A_k y_k^n + f_{k+1}^n$$

to the limit as $n \to \infty$ to show that

$$y_{k+1} = A_k y_k + f_{k+1}.$$

Hence, $y = T(f)$, and the graph of T is closed. □
Lemma 2.1.1 implies that there exists a constant $r > 0$ such that

$$\|T(f)\| \leq r\|f\|, \quad f \in \mathscr{B}. \tag{2.9}$$

Without loss of generality, we assume that

$$rN \geq 1, \tag{2.10}$$

where N is the constant in (2.1).
Denote

$$X(k) = \begin{cases} F(k,0), & k > 0; \\ \mathrm{Id}, & k = 0; \\ F(0,-k), & k < 0. \end{cases}$$

Straightforward calculations show that the formula

$$y_k = \sum_{u=0}^{k} X(k)PX(-u)f_u - \sum_{u=k+1}^{\infty} X(k)(\mathrm{Id} - P)X(-u)f_u \tag{2.11}$$

represents a solution of Eq. (2.3) provided that the series in the second summand converges.
We can obtain a shorter variant of formula (2.11) by introducing the "Green function"

$$G(k,u) = \begin{cases} X(k)PX(-u), & 0 \leq u \leq k; \\ -X(k)(\mathrm{Id} - P)X(-u), & 0 \leq k < u. \end{cases}$$

Then formula (2.11) becomes

$$y_k = \sum_{u=0}^{\infty} G(k,u)f_u. \tag{2.12}$$

Lemma 2.1.2 *Let $k_0, k_1, k \in I$ and let $\xi \in \mathbb{R}^n$ be a nonzero vector with $|\xi| \le 1$. Then*

$$|X(k)P\xi| \sum_{u=k_0}^{k} |X(u)\xi|^{-1} \le r, \quad 0 \le k_0 \le k, \tag{2.13}$$

and

$$|X(k)(Id - P)\xi| \sum_{u=k}^{k_1} |X(u)\xi|^{-1} \le 2rN, \quad 0 \le k \le k_1. \tag{2.14}$$

Proof Without loss of generality, we may take $f_0 = 0$. Fix $l_0, l_1 \in I$ such that $l_0 \le l_1$. Take a sequence f with $f_i = 0,\ i > l_1$. Then formula (2.12) takes the form

$$y_l = \sum_{u=0}^{l_1} G(l, u)f_u.$$

For $l \ge l_1$, all the indices u in this sum do not exceed l_1, and we apply the first line in the definition of G. Thus,

$$y_l = X(l)P \sum_{u=0}^{l_1} X(-u)f_u.$$

Hence, if $l \ge l_1$, then y_l is the image under $X(l)$ of a vector from V_1 that does not depend on l. It follows that the sequence y (with the exception of a finite number of entries) is a solution of Eq. (2.2) with initial value from V_1. Hence, $y \in \mathscr{B}$. Since $f_0 = 0$,

$$y_0 = -(\mathrm{Id} - P) \sum_{u=0}^{l_1} X(-u)f_u \in V_2.$$

Thus, $y = T(f)$, and $\| y\| \le r\|f\|$.
Now we specify the choice of f. Let $x_i = X(i)\xi$; since $\xi \ne 0$, $x \ne 0$ as well. Set

$$f_i = \begin{cases} 0, & i < l_0; \\ x_i/|x_i|, & l_0 \le i \le l_1; \\ 0, & i > l_1. \end{cases}$$

Since $\|f\| = 1$, inequality (2.9) implies that

$$\left| \sum_{u=l_0}^{l_1} G(k, u)x_i/|x_i| \right| = |\, y_l| \le r. \tag{2.15}$$

We take $l = l_1 = k$ and $l_0 = k_0$ in (2.15) and conclude that

$$r \geq \left| \sum_{u=k_0}^{k} G(k,u)x_u/|x_u| \right| = \left| \sum_{u=k_0}^{k} X(k)PX(-u)X(u)\xi/|X(u)\xi| \right| =$$

$$= |X(k)P\xi| \sum_{u=k_0}^{k} |X(u)\xi|^{-1},$$

which is precisely inequality (2.13).

We prove inequality (2.14) using a similar reasoning.

First we consider $0 < k \leq k_1$. We take $l = k-1$, $l_0 = k$, and $l_1 = k_1$ in (2.15) and get the estimates

$$r \geq \left| \sum_{u=k}^{k_1} G(k,u)x_u/|x_u| \right| = \left| \sum_{u=k}^{k_1} X(k-1)(\mathrm{Id}-P)X(-u)X(u)\xi/|X(u)\xi| \right| =$$

$$= |X(k-1)(\mathrm{Id}-P)\xi| \sum_{u=k}^{k_1} |X(u)\xi|^{-1} = |A_{k-1}^{-1}X(k)(\mathrm{Id}-P)\xi| \sum_{u=k}^{k_1} |X(u)\xi|^{-1} \geq$$

$$\geq \|A_{k-1}\|^{-1} |X(k)(\mathrm{Id}-P)\xi| \sum_{u=k}^{k_1} |X(u)\xi|^{-1}.$$

Applying inequality (2.1), we see that in this case,

$$|X(k)(\mathrm{Id}-P)\xi| \sum_{u=k}^{k_1} |X(u)\xi|^{-1} \leq rN.$$

Now we consider $0 = k < k_1$ and apply the previous estimate with $k = 1$:

$$|X(0)(\mathrm{Id}-P)\xi| \sum_{u=0}^{k_1} |X(u)\xi|^{-1} = |X(0)(\mathrm{Id}-P)\xi| \sum_{u=1}^{k_1} |X(u)\xi|^{-1} + |(\mathrm{Id}-P)\xi| \leq$$

$$\leq \|A_0\|^{-1} |X(1)(\mathrm{Id}-P)\xi| \sum_{u=1}^{k_1} |X(u)\xi|^{-1} + 1 \leq rN + 1 \leq 2rN$$

(recall that $|\xi| \leq 1$ and $rN \geq 1$).

For $k = k_1 = 0$, our inequality is trivial. □

Lemma 2.1.3 *Let $k_0, k_1, k, s \in I$ and let $\xi \in \mathbb{R}^n$ be a unit vector. Denote*

$$\mu = 1 - (2rN)^{-1}.$$

Then the following inequalities are satisfied:
if $P\xi \neq 0$, then

$$\sum_{u=k_0}^{s} |X(u)P\xi|^{-1} \leq \mu^{k-s} \sum_{u=k_0}^{k} |X(u)P\xi|^{-1}, \quad k_0 \leq s \leq k; \tag{2.16}$$

if $(Id - P)\xi \neq 0$, then

$$\sum_{u=s}^{k_1} |X(u)(Id - P)\xi|^{-1} \leq \mu^{s-k} \sum_{u=k}^{k_1} |X(u)(Id - P)\xi|^{-1}, \quad k \leq s \leq k_1. \tag{2.17}$$

Proof Denote

$$\phi_i = \sum_{u=k_0}^{i} |X(u)P\xi|^{-1}, \quad i \geq k_0,$$

and

$$\psi_i = \sum_{u=i}^{k_1} |X(u)(\mathrm{Id} - P)\xi|^{-1}, \quad i \leq k_1.$$

Let us prove inequality (2.16). Since $P\xi \neq 0$, $\phi_i > 0$. Clearly, $\phi_i - \phi_{i-1} = |X(i)P\xi|^{-1}$. Replacing ξ by $P\xi$ (and noting that $|P\xi| \leq 1$) in (2.13), we see that

$$\frac{\phi_i}{\phi_i - \phi_{i-1}} \leq r \leq 2rN.$$

Hence,

$$(2rN)^{-1} \leq \frac{\phi_i - \phi_{i-1}}{\phi_i} = 1 - \frac{\phi_{i-1}}{\phi_i},$$

and

$$\phi_{i-1} \leq (1 - (2rN)^{-1})\phi_i.$$

Iterating this inequality, we conclude that

$$\phi_s \leq (1 - (2rN)^{-1})^{k-s}\phi_k, \quad k \geq s.$$

We prove inequality (2.17) similarly. We note that $\psi_i > 0$ and that $\psi_i - \psi_{i+1} = |X(i)(\mathrm{Id} - P)\xi|^{-1}$. After that, we replace ξ by $(\mathrm{Id} - P)\xi$ in (2.14) and show that

$$\psi_{i+1} \leq (1 - (2rN)^{-1})\psi_i.$$

Iterating this inequality, we get (2.17). □

Now we prove that the sequence $\mathscr{A}$ is hyperbolic.

Lemma 2.1.4 *The following inequalities are satisfied:*

$$\|X(k)PX(-s)\| \leq r^2\mu^{k-s}, \quad 0 \leq s \leq k,$$

and

$$\|X(k)(Id - P)X(-s)\| \leq 2r^2N^2\mu^{s-k}, \quad 0 \leq k \leq s.$$

Proof Fix a natural s and a unit vector ξ. Define a sequence $y = \{y_k\}$ by

$$y_k = \begin{cases} -X(k)(\mathrm{Id} - P)X(-s)\xi, & 0 \leq k < s; \\ X(k)PX(-s)\xi, & k \geq s. \end{cases}$$

The sequence y coincides (up to a finite number of terms) with a solution of Eq. (2.2) with initial point from V_1; hence, $y \in \mathscr{B}$.

Now we define a sequence f by

$$f_k = \begin{cases} 0, & k \neq s; \\ \xi, & k = s. \end{cases}$$

It is easily seen that the above sequence y is a solution of Eq. (2.3) with inhomogeneity f. Hence, $y = T(f)$, and $\| y\| \leq r$.

The definition of y implies that

$$|X(k)PX(-s)\xi| = |y_k| \leq r, \quad 0 \leq s \leq k.$$

Since ξ is an arbitrary unit vector, $\|X(k)PX(-s)\| \leq r$ for $0 \leq s \leq k$.

We replace ξ by the solution of the equation $x_s = X(s)\xi$ to show that

$$|X(k)P\xi| = |X(k)PX(-s)x_s| \leq r|x_s|, \quad 0 \leq s \leq k. \tag{2.18}$$

Using inequalities (2.13), (2.16) with $k_0 = s$, and (2.18) with $k = s$, we see that

$$|X(k)PX(-s)x_s| = |X(k)P\xi| \leq r\left(\sum_{u=s}^{k} |X(u)P\xi|^{-1}\right)^{-1} \leq$$

$$\leq r\left(\mu^{-(k-s)}|X(s)P\xi|^{-1}\right)^{-1} = r\mu^{k-s}|X(s)P\xi| \leq r^2\mu^{k-s}|x_s|.$$

If $P\xi = 0$, then the resulting estimate is obvious. Since $x_s = X(s)\xi$ and $X(s)$ is an isomorphism, we get the following estimate for the operator norm:

$$\|X(k)PX(-s)\| \le r^2\mu^{k-s}, \quad 1 \le s \le k.$$

In this reasoning, we have used inequality (2.18) with $s = k$. It is also true for $s = k = 0$ since $\|P\| \le 1$. Therefore, the first estimate of our lemma is proved for $0 \le s \le k$.

The proof of the second estimate is quite similar. The only difference is as follows. We cannot use an analog of (2.18) with $k = s$ since $k \ne s$ in the definition of the sequence y. The following inequality is proved by the same reasoning as above:

$$|X(k)(\mathrm{Id} - P)\xi| = |X(k)(\mathrm{Id} - P)X(-s)x_s| \le r|x_s|, \quad s > k.$$

In the case $k = s - 1$, we write

$$|X(s)(\mathrm{Id} - P)\xi| = |A_{s-1}X(s-1)(\mathrm{Id} - P)X(-s)x_s| \le$$

$$\le \|A_{s-1}\||X(s-1)(\mathrm{Id} - P)X(-s)x_s| \le rN|x_s|,$$

and then repeat the reasoning of the first case. □

Lemma 2.1.4 shows that if we take constants $C_0 = r^2N$ and $\lambda = \mu$ and projections

$$P_k = X(k)PX(-k) \quad \text{and} \quad Q_k = X(k)(\mathrm{Id} - P)X(-k),$$

then the operators $F(k, l)$ generated by the sequence $\mathscr{A}$ satisfy estimates (2.6) and (2.7) with $C = C_0$ and λ. Clearly, relations (2.4) and (2.5) are valid.

Thus, to show that $\mathscr{A}$ is hyperbolic on I, it remains to prove the following statement.

Lemma 2.1.5 *There exists a constant $C = C(N, C_0, \lambda) \ge C_0$ such that inequalities (2.8) are fulfilled.*

Proof Let L_1 and L_2 be two linear subspaces of $\mathbb{R}^n$. Introduce the value

$$\angle(L_1, L_2) = \min |v_1 - v_2|,$$

where the minimum is taken over all pairs of unit vectors $v_1 \in L_1$, $v_2 \in L_2$.

We claim that there exists a constant $C_1 = C_1(N, C_0, \lambda)$ such that

$$\angle(S_k, U_k) \ge C_1, \quad k \in I. \tag{2.19}$$

Fix an index $k \in I$, take unit vectors $v_1 \in S_k$ and $v_2 \in U_k$ for which $\angle(S_k, U_k) = |v_1 - v_2|$, and denote

$$\alpha_l = |\, F(l,k)(v_1 - v_2)|, \quad l \geq k.$$

Inequalities (2.6) and (2.7) imply that

$$\alpha_l \geq |\, F(l,k)v_2| - |\, F(l,k)v_1| \geq \lambda^{k-l}/C_0 - C_0\lambda^{l-k}.$$

Hence, there exists a constant $m = m(C_0, \lambda)$ such that

$$\alpha_{k+m} \geq 1.$$

At the same time, it follows from (2.1) that

$$\alpha_{k+m} \leq N^m \alpha_k.$$

Combining the above two inequalities, we see that

$$\angle(S_k, U_k) = \alpha_k \geq C_1(N, C_0, \lambda) := N^{-m(C_0,\lambda)},$$

which proves (2.19).

Clearly, if v_1 and v_2 are two unit vectors, then the usual angle $\langle v_1, v_2 \rangle$ satisfies the relation

$$|v_1 - v_2| = 2\sin(\langle v_1, v_2 \rangle/2),$$

and we see that estimate (2.19) implies the existence of $\beta = \beta(N, C_0, \lambda)$ such that if γ is the usual angle between S_k and U_k, then

$$\sin(\gamma) \geq \beta.$$

Now we take an arbitrary unit vector $v \in \mathbb{R}^n$ and denote $v_s = P_k v$. If γ_s is the angle between v and v_s, then the sine law implies that

$$\frac{|v|}{\sin(\gamma)} = \frac{|v_s|}{\sin(\gamma_0)} \geq |v_s|,$$

and we conclude that

$$|v_s| = |\, P_k v| \leq 1/\beta,$$

which implies that

$$\| P_k \| \le C = \max(C_0, 1/\beta).$$

A similar estimate holds for $\|Q_k\|$. □

As we said above, the following statement holds.

Lemma 2.1.6 *A hyperbolic sequence $\mathscr{A}$ has the Perron property on I.*

Proof Assume that the sequence $\mathscr{A}$ has properties stated in relations (2.4)–(2.8). Take a sequence

$$f = \{f_k \in \mathbb{R}^n : k \in I\}$$

such that $\|f\| = \nu < \infty$ and consider the sequence y defined by formula (2.11).

Then

$$|X(k)PX(-u)f_u| \le C\lambda^{k-u}\nu, \quad 0 \le u \le k,$$

and

$$|X(k)(\mathrm{Id} - P)X(-u)f_u| \le C\lambda^{u-k}\nu, \quad k+1 \le u < \infty,$$

which implies that the second term in (2.11) is a convergent series (hence, the sequence y is a solution of (2.3)) and the estimate

$$\| y\| \le C(1 + \lambda + \lambda^2 \dots)\nu + C(\lambda + \lambda^2 \dots)\nu = \frac{1+\lambda}{1-\lambda}C\nu$$

holds. □

Now we pass to the Pliss theorem.

This time, $I = \mathbb{Z}$, and we denote $I_+ = \{k \in \mathbb{Z} : k \ge 0\}$ and $I_- = \{k \in \mathbb{Z} : k \le 0\}$.

Now $\mathscr{A}$ is a sequence of linear isomorphisms

$$A_k : \mathbb{R}^n \to \mathbb{R}^n, \quad k \in I = \mathbb{Z}.$$

It is again assumed that an analog of inequalities (2.1) holds, and we consider difference equations (2.2) and (2.3).

The Perron property of (2.2) on $\mathbb{Z}$ is defined literally as in the case of $I = \{k \in \mathbb{Z} : k \ge 0\}$.

It follows from the Maizel' theorem and its obvious analog for the case of $I = \{k \in \mathbb{Z} : k \le 0\}$ that the sequence $\mathscr{A}$ is hyperbolic on both I_+ and I_- (the definition of hyperbolicity in the case of I_- is literally the same).

Without loss of generality, we assume that C and λ are the same for the hyperbolicity on I_+ and I_- and denote by $S_k^+, U_k^+, k \in I_+$, and $S_k^-, U_k^-, k \in I_-$, the corresponding subspaces of $\mathbb{R}^n$.

Theorem 2.1.2 (Pliss) *If $\mathscr{A}$ has the Perron property on $I = \mathbb{Z}$, then the subspaces U_0^- and S_0^+ are transverse.*

Remark 2.1.2 In fact, Pliss proved in [74] that the transversality of U_0^- and S_0^+ is equivalent to the Perron property of $\mathscr{A}$ on $I = \mathbb{Z}$, but we need only the implication stated above.

Remark 2.1.3 Note that there exist sequences $\mathscr{A}$ that are separately hyperbolic on I_+ and I_- for which the subspaces U_0^- and S_0^+ are transverse and such that these sequences are not hyperbolic on $I = \mathbb{Z}$. It is easy to construct such a sequence with $S_k^+ = \mathbb{R}^n, U_k^+ = \{0\}, k \in I_+$, and $S_k^- = \{0\}, U_k^- = \mathbb{R}^n, k \in I_-$ (we leave details to the reader).

Proof To get a contradiction, assume that the subspaces U_0^- and S_0^+ are not transverse. Then there exists a vector $x \in \mathbb{R}^n$ such that

$$x \neq y_1 + y_2 \tag{2.20}$$

for any $y_1 \in U_0^-$ and $y_2 \in S_0^+$.

Since the subspaces U_0^+ and S_0^+ are complementary (see (2.4)), we can represent

$$x = \xi + \eta, \quad \xi \in S_0^+, \ \eta \in U_0^+.$$

Then it follows from (2.20) that

$$\eta \neq z_1 + z_2 \tag{2.21}$$

for any $z_1 \in S_0^+$ and $z_2 \in U_0^-$. We may assume that $|\eta| = 1$.

Consider the sequence

$$a_k = \begin{cases} 0, \ k \leq 0; \\ 1, \ k > 0. \end{cases}$$

Since $\eta \neq 0$ in (2.21), $X(k)\eta \neq 0$ for $k \in I$. Define a sequence $f = \{f_k, \ k \in I\}$ by

$$f_k = \frac{X(k)\eta}{|X(k)\eta|} a_k, \quad k \in I. \tag{2.22}$$

Clearly, $\|f\| = 1$. We claim that the corresponding Eq. (2.3) does not have bounded solutions.

Consider the sequence

$$\phi_k = - \sum_{u=k+1}^{\infty} X(k)(\mathrm{Id} - P)X(-u)f_u, \quad k \geq 0.$$

In this formula, P is the projection defined for Eq. (2.2).

The sequence $\{\phi_k\}$ is bounded for $k \geq 0$. Indeed, $f_u \in U_u^+$ for $u \geq 0$; hence,

$$|\phi_k| = \left| \sum_{u=k+1}^{\infty} X(k)(\mathrm{Id} - P)X(-u)f_u \right| \leq$$

$$\leq \sum_{u=k+1}^{\infty} C\lambda^{u-k} = C\frac{\lambda}{1-\lambda}.$$

We know that since the series defining ϕ_k is convergent, the sequence $\{\phi_k\}$ is a solution of the homogeneous equation (2.2) for $k \geq 0$.

Clearly,

$$\phi_0 = -\sum_{u=1}^{\infty} (\mathrm{Id} - P)X(-u)f_u = -\sum_{u=1}^{\infty} \frac{\eta}{|X(u)\eta|} = \nu\eta,$$

where

$$\nu = -\sum_{u=1}^{\infty} \frac{1}{|X(u)\eta|}.$$

Deriving these relations, we take into account the definition of f and the equality $(\mathrm{Id} - P)\eta = \eta$. In addition, the value ν is finite since

$$\frac{1}{|X(k)\eta|} \leq C\lambda^k, \quad k \geq 0,$$

due to inequalities (2.7).

It follows from (2.21) that

$$\phi_0 \neq y_1 + y_2 \tag{2.23}$$

for any $y_1 \in S_0^+$ and $y_2 \in U_0^-$.

Now let us assume that Eq. (2.2) has a solution $\psi = \{\psi_k\}$ that is bounded on $I = \mathbb{Z}$. Then $\psi_0 \in U_0^-$.

On the other hand,

$$\psi_k = X(k)(\psi_0 - \phi_0) + \phi_0.$$

Since ϕ_k are bounded for $k \geq 0$, ψ_k can be bounded for $k \geq 0$ only if

$$X(k)(\psi_0 - \phi_0)$$

are bounded for $k \geq 0$, which implies that

$$\psi_0 - \phi_0 \in S_0^+.$$

Set

$$y_1 = \phi_0 - \psi_0 \in S_0^+ \quad \text{and} \quad y_2 = \psi_0 \in U_0^-.$$

Then $\phi_0 = y_1 + y_2$, and we get a contradiction with (2.23). □

Remark 2.1.4 We will apply the Maizel' and Pliss theorems proved in this section in a slightly different situation.

We consider a diffeomorphism f of a smooth closed manifold M, fix a point $x \in M$ and the trajectory $\{x_k = f^k(x) : k \in \mathbb{Z}\}$ of this point and define linear isomorphisms

$$A_k = Df(x_k) : T_{x_k}M \to T_{x_{k+1}}M.$$

To the sequence $\mathscr{A} = \{A_k\}$ we assign difference equations

$$v_{k+1} = A_k v_k, \quad v_k \in T_{x_k}M,$$

and

$$v_{k+1} = A_k v_k + f_{k+1}, \quad v_k \in T_{x_k}M, \, f_{k+1} \in T_{x_{k+1}}M.$$

Clearly, these difference equations are completely similar to Eqs. (2.2) and (2.3), and analogs of the Maizel' and Pliss theorems are valid for them.

Historical Remarks Theorem 2.1.1 was proved by A. D. Maizel' in [38]. See also the classical W. A. Coppel's book [13].

The Pliss theorem (Theorem 2.1.2) was published in [74]. Later, it was generalized by many authors; let us mention, for example, K. Palmer [55] who studied Fredholm properties of the corresponding operators.

2.2 Mañé Theorem

In this section, we prove Theorem 1.3.7.

Remark 2.2.1 In several papers, the analytic strong transversality condition is formulated in the following form, which is obviously stronger than the condition formulated in Definition 1.3.11: it is assumed that

$$\tilde{B}^+(x) + \tilde{B}^-(x) = T_xM, \quad x \in M,$$

where the subspaces $\tilde{B}^+(x)$ and $\tilde{B}^-(x)$ are defined by the equalities

$$\tilde{B}^+(x) = \left\{ v \in T_xM : \lim_{k \to \infty} \left| Df^k(x)v \right| = 0 \right\}$$

and

$$\tilde{B}^-(x) = \left\{ v \in T_xM : \lim_{k \to -\infty} \left| Df^k(x)v \right| = 0 \right\}.$$

In fact, it is easily seen from our proof below that the structural stability of f implies this form of the analytic strong transversality condition as well, so that both conditions are equivalent.

The main part of our proof of Theorem 1.3.7 is contained in the following statement.

Theorem 2.2.1 *The analytic strong transversality condition implies Axiom A.*

First we prove that the analytic strong transversality condition implies the hyperbolicity of the nonwandering set Ω.

We assign to a diffeomorphism $f : M \to M$ the mapping $\pi : TM \to TM$ (where TM is the tangent bundle of M) which maps a pair $(x, v) \in TM$ (where $x \in M$ and $v \in T_xM$) to the pair $(f(x), Df(x)v)$.

A subbundle Y of TM is a set of pairs (x, Y_x), where $x \in M$ and Y_x is a linear subspace of T_xM.

Definition 2.2.1 A subbundle Y is called *π-invariant* if

$$Df(x)Y_x = Y_{f(x)} \quad \text{for} \quad x \in M.$$

Assuming that f satisfies the analytic strong transversality condition, we define two subbundles B^+ and B^- of TM by setting

$$B_x^+ = B^+(x) \quad \text{and} \quad B_x^- = B^-(x) \quad \text{for} \quad x \in M.$$

Since

$$\lim_{k\to\infty}\inf \left|Df^k(x)v\right| = 0$$

if and only if

$$\lim_{k\to\infty}\inf \left|Df^k(f(x))Df(x)v\right| = 0,$$

the subbundle B^+ is π-invariant. A similar reasoning shows that the subbundle B^- is π-invariant as well.

The main object in the proof is the mapping π^*, dual to the mapping π.

Denote by $<,>$ the scalar product in T_xM. Let $D^*f(x) : T_{f(x)}M \to T_xM$ be defined as follows:

$$< \xi, Df(x)v >=< D^*f(x)\xi, v >$$

for all $v \in T_xM$ and $\xi \in T_{f(x)}M$ (thus, $D^*f(x)$ is the adjoint of $Df(x)$). We define π^* as follows: a pair $(f(x), \xi)$, $\xi \in T_{f(x)}M$, is mapped to

$$\pi^*\left(f(x), \xi\right) = \left(x, D^*f(x)\xi\right).$$

If $p : TM \to M$ is the projection to the first coordinate (i.e., $p(x, v) = x$), then $p(\pi(x, v)) = f(x)$ (in this case, one says that π covers f); since $p(\pi^*(x, v)) = f^{-1}(x)$, π^* covers f^{-1}.

Clearly, the definition of π^* implies the following statement.

Lemma 2.2.1

$$(\pi^*)^* = \pi.$$

If Y is a subbundle of TM, we define the orthogonal subbundle $Y^\perp$ as follows:

$$Y_x^\perp = \{\xi : < \xi, v >= 0 \quad \text{for all} \quad v \in Y_x\}, \quad x \in M.$$

Lemma 2.2.2 *If a subbundle Y is π-invariant, then $Y^\perp$ is π^*-invariant.*

Proof Consider vectors $\xi \in Y_{f(x)}^\perp$ and $D^*f(x)\xi \in T_xM$. If $v \in Y_x$, then

$$< v, D^*f(x)\xi >=< \xi, Df(x)v >= 0$$

since $Df(x)v \in Y_{f(x)}$, which means that $D^*f(x)\xi \in Y_x^\perp$. □

We call two subbundles Y^1 and Y^2 complementary if

$$Y_x^1 \oplus Y_x^2 = T_xM \quad \text{for any} \quad x \in M. \tag{2.24}$$

Lemma 2.2.3 *If Y^1 and Y^2 are complementary subbundles that are π-invariant, then $\left(Y^1\right)^{\perp}$ and $\left(Y^2\right)^{\perp}$ are complementary subbundles that are π^*-invariant.*

Proof The subbundles $\left(Y^1\right)^{\perp}$ and $\left(Y^2\right)^{\perp}$ are π^*-invariant by Lemma 2.2.2. If $\dim Y^1_x = k$, then equality (2.24) implies that $\dim Y^2_x = n - k$. Clearly,

$$\dim\left(Y^1\right)^{\perp}_x = n - k \quad \text{and} \quad \dim\left(Y^2\right)^{\perp}_x = k. \tag{2.25}$$

Consider a vector $\xi \in \left(Y^1\right)^{\perp}_x \cap \left(Y^2\right)^{\perp}_x$. Due to (2.24), any vector $v \in T_xM$ is representable as

$$v = v_1 + v_2, \quad v_1 \in Y^1_x, \; v_2 \in Y^2_x.$$

Then $< \xi, v >=< \xi, v_1 > + < \xi, v_2 >= 0$. Since v is arbitrary, $\xi = 0$. The equality

$$\left(Y^1\right)^{\perp}_x \cap \left(Y^2\right)^{\perp}_x = \{0\}$$

and (2.25) imply the statement of our lemma. □

Let $M_0 \subset M$ be a hyperbolic set of f. Then S and U defined by $S_x = S(x)$ and $U_x = U(x)$ for $x \in M_0$ are two complementary π-invariant subbundles on M_0 such that inequalities (HSD2.3) and (HSD2.4) hold (see Definition 1.3.1). In this case, we say that M_0 is hyperbolic with respect to π with subbundles S and U and constants C and λ.

Lemma 2.2.4 *If a set M_0 is hyperbolic with respect to π with subbundles S and U and constants C and λ, then M_0 is hyperbolic with respect to π^* with subbundles $U^{\perp}$ and $S^{\perp}$ and the same constants C and λ.*

Proof If A and B are linear operators, then $(AB)^* = B^*A^*$; hence,

$$\left(Df(f(x))Df(x)\right)^* = D^*f(x)D^*f(f(x)).$$

If we take $v \in T_xM$ and $\xi \in T_{f^2(x)}M$, then

$$< Df^2(x)v, \xi >=< Df(f(x))Df(x)v, \xi >=$$

$$=< Df(x)v, D^*f(f(x))\xi >=< v, D^*f(x)D^*f(f(x))\xi >=< v, D^*f^2(x)\xi > .$$

Applying induction, it is easy to show that

$$< v, D^*f^k(x)\xi >=< \xi, Df^k(x)v >, \quad k \in \mathbb{Z}, \tag{2.26}$$

for $v \in T_xM$ and $\xi \in T_{f^k(x)}M$, where

$$D^*f^k(x) = D^*f(x)D^*f(f(x))\dots D^*f(f^{k-1}(x))$$

and

$$Df^k(x) = Df(f^{k-1}(x))Df(f^{k-2}(x))\dots Df(x).$$

By Lemma 2.2.3, the subbundles $S^\perp$ and $U^\perp$ are complementary and π^*-invariant.

Fix $k \geq 0$ and a vector $\xi \in \left(U_{f^k(x)}\right)^\perp$. Then $D^*f^k(x)\xi \in T_xM$. The obvious equality

$$|\eta| = \max_{|v|=1} < \eta, v >, \quad \eta, v \in T_xM,$$

implies that

$$\left|D^*f^k(x)\xi\right| = \max_{|v|=1} < v, D^*f^k(x)\xi > .$$

Represent $v = v_1 + v_2$, where $v_1 \in S_x$ and $v_2 \in U_x$.

Since $U^\perp$ is π^*-invariant,

$$D^*f^k(x)\xi \in (U_x)^\perp,$$

and $< v_2, D^*f^k(x)\xi >= 0$. It follows that

$$\left|D^*f^k(x)\xi\right| = \max_{|v_1|=1} < v_1, D^*f^k(x)\xi >= \max_{|v_1|=1} < \xi, Df^k(x)v_1 >\leq C\lambda^k|\xi|.$$

In the last inequality, we used inequality (HSD2.3) and the obvious relation

$$< \xi, v >\leq |\xi||v|.$$

A similar reasoning shows that

$$\left|D^*f^{-k}(x)\xi\right| \leq C\lambda^{-k}|\xi|$$

for $\xi \in \left(S_{f^k(x)}\right)^\perp$ and $k \leq 0$. □

Now we prove that the analytic strong transversality condition implies that, in a sense, π^* does not have nontrivial bounded trajectories. Fix a point $(x, v) \in TM$ and define the sequence $(x_k, v_k) = (\pi^*)^k(x, v)$.

Lemma 2.2.5 *If*

$$\sup_{k\in\mathbb{Z}} |v_k| < \infty, \tag{2.27}$$

then $v = 0$.

Proof The obvious equalities

$$x = f^{-k}\left(f^k(x)\right) \quad \text{and} \quad u = Df^{-k}\left(f^k(x)\right) Df^k(x)u$$

which are valid for all $x \in M$, $u \in T_xM$, and $k \in \mathbb{Z}$ imply that

$$< \xi, u >=< \xi, Df^{-k}\left(f^k(x)\right) Df^k(x)u >=< D^*f^{-k}\left(f^k(x)\right)\xi, Df^k(x)u >$$

for all $\xi, u \in T_xM$ and k.

Assume that a point (x, v) satisfies condition (2.27).

By the analytic strong transversality condition, we can represent any vector $\xi \in T_xM$ in the form $\xi = \xi_1 + \xi_2$ for which there exist sequences $l_n \to \infty$ and $m_n \to -\infty$ as $n \to \infty$ such that

$$\left|Df^{l_n}(x)\xi_1\right| \to 0 \quad \text{and} \quad \left|Df^{m_n}(x)\xi_2\right| \to 0, \quad n \to \infty.$$

Let us write

$$< v, \xi >=< v, \xi_1 + \xi_2 >=< v, Df^{-l_n}\left(f^{l_n}(x)\right) Df^{l_n}(x)\xi_1 > +$$

$$+ < v, Df^{-m_n}\left(f^{m_n}(x)\right) Df^{m_n}(x)\xi_1 >=$$

$$=< D^*f^{-l_n}\left(f^{l_n}(x)\right) v, Df^{l_n}(x)\xi_1 > + < D^*f^{-m_n}\left(f^{m_n}(x)\right) v, Df^{m_n}(x)\xi_2 > . \tag{2.28}$$

By condition (2.27), both values $\left|D^*f^{-l_n}(f^{l_n}(x))v\right|$ and $\left|D^*f^{-m_n}(f^{m_n}(x))v\right|$ are bounded; hence, both terms in (2.28) tend to 0 as $n \to \infty$. Thus, $< \xi, v >= 0$ for any ξ, which means that $v = 0$. □

To simplify notation, let us denote π^* by ρ and write

$$\rho(x, v) = (\phi(x), \Phi(x)v),$$

so that $\phi(x) = f^{-1}(x)$ and $\Phi(x)$ is the linear mapping $T_xM \to T_{\phi(x)}M$, $\Phi(x) = D^*f(x)$. Let

$$F(0, x) = \text{Id},$$

$$F(k, x) = \Phi(\phi^{k-1}(x)) \cdots \Phi(x), \quad k > 0,$$

and

$$F(-k,x) = \Phi^{-1}(\phi^{1-k}(x)) \cdots \Phi^{-1}(x), \quad k > 0.$$

Obviously, the mapping ρ is continuous. By Lemma 2.2.5, it satisfies the following *Condition B*: If

$$\sup_{k \in \mathbb{Z}} |F(k,x)v| < \infty$$

for some $(x, v) \in TM$, then $v = 0$.

Let us define the following two subbundles in TM: $V = \{(x, V_x)\}$ and $W = \{(x, W_x)\}$. We agree that

- $v \in T_xM$ belongs to V_x if $|F(k,x)v| \to 0$ as $k \to \infty$
  and
- $v \in T_xM$ belongs to W_x if $|F(k,x)v| \to 0$ as $k \to -\infty$.

Clearly, the subbundles V and W are ρ-invariant.

Lemma 2.2.6 *Let a sequence $(x_m, v_m) \in TM$ be such that*

(1) $(x_m, v_m) \to (x, v)$ as $m \to \infty$;
(2) there exists a number $L > 0$ and a sequence $k_m \to \infty$ as $m \to \infty$ such that

$$|F(k, x_m)v_m| \le L, \quad 0 \le k \le k_m. \tag{2.29}$$

Then $(x, v) \in V$.

Proof Fix an arbitrary $l \ge 0$. There exists an m_0 such that $k_m > l$ for $m \ge m_0$. Then it follows from (2.29) that

$$|F(l, x_m)v_m| \le L. \tag{2.30}$$

Since $F(l, y)w$ is continuous in y and w, we may pass to the limit in (2.30) as $m \to \infty$; thus,

$$|F(l,x)v| \le L.$$

Since l is arbitrary, this means that

$$|F(k,x)v| \le L, \quad k \ge 0. \tag{2.31}$$

Let (x_0, v_0) be a limit point of the sequence $\big(\phi^k(x), F(k,x)v\big)$, i.e., the limit of the sequence

$$(\phi^{t_m}(x), F(t_m, x)v) \tag{2.32}$$

for some sequence $t_m \to \infty$.

Take an arbitrary $k \in \mathbb{Z}$. Since

$$\phi^{t_m}(x) \to x_0 \quad \text{and} \quad F(t_m, x)v \to v_0, \quad m \to \infty,$$

$$\phi^{k+t_m}(x) \to \phi^k(x_0) \quad \text{and} \quad F(k + t_m, x)v \to F(k, x_0)v_0, \quad m \to \infty. \tag{2.33}$$

For large m, $k + t_m > 0$, and it follows from (2.31) and the second relation in (2.33) that

$$|F(k, x_0)v_0| \leq L. \tag{2.34}$$

Since (2.34) is valid for any $k \in \mathbb{Z}$, Condition B implies that $v_0 = 0$. Thus, in any convergent sequence of the form (2.32) with $t_m \to \infty$,

$$|F(t_m, x)v| \to 0,$$

which means that $(x, v) \in V$. □

Remark 2.2.2 A similar reasoning shows that if we take $k_m \to -\infty$ and $k_m \leq k \leq 0$ in condition (2) of Lemma 2.2.6, then $(x, v) \in W$. In what follows, we do not make such comments and only consider the case of the subbundle V.

Define the set

$$A = \{(x, v) \in TM : |F(k, x)v| \leq 1 \quad \text{for} \quad k \geq 1\}.$$

Clearly, the set A is positively ρ-invariant, i.e., if $(x, v) \in A$ and $k \geq 0$, then $(\phi^k(x), F(k, x)v) \in A$.

Let us say that a set $C = \{(x, v) \in TM\}$ is bounded if

$$\sup_{(x,v)\in C} |v| < \infty.$$

Since the manifold M is compact, any closed and bounded subset C of TM is (sequentially) compact, i.e., any sequence in C has a convergent subsequence, and the limit of this subsequence belongs to C.

Lemma 2.2.7 *The set A is a compact subset of V.*

Proof It was shown in the proof of Lemma 2.2.6 that inequality (2.31) implies the inclusion $(x, v) \in V$; thus, $A \subset V$. Since $F(0, x)v = v$, A is bounded. Consider a sequence $(x_m, v_m) \in A$ such that $(x_m, v_m) \to (x, v)$, $m \to \infty$. For any fixed $k \geq 0$,

$$|F(k, x)v| = \lim_{m\to\infty} |F(k, x_m)v_m| \leq 1.$$

Hence, $(x, v) \in A$, and A is closed. □

Lemma 2.2.8 *For any $\mu > 0$ there exists a $K > 0$ such that if $(x, v) \in A$, then*

$$|F(k,x)v| < \mu, \quad k \geq K. \tag{2.35}$$

Proof Assuming the converse, let us find sequences $(x_m, v_m) \in A$ and $k_m \to \infty$ and a number $\mu > 0$ such that

$$|F(k_m, x_m)v| \geq \mu. \tag{2.36}$$

Since A is positively ρ-invariant,

$$\left(\phi^{k_m}(x_m), F(k_m, x_m)v_m\right) \in A;$$

since A is compact, the above sequence has a convergent subsequence. Assume, for definiteness, that

$$\left(\phi^{k_m}(x_m), F(k_m, x_m)v_m\right) \to (x, v).$$

Then it follows from (2.36) that $|v| \geq \mu$. Fix a number $k \in \mathbb{Z}$. Since $k + k_m > 0$ for large m,

$$\left(\phi^{k+k_m}(x_m), F(k + k_m, x_m)v_m\right) \to \left(\phi^k(x), F(k,x)v\right), \quad m \to \infty,$$

and

$$|F(k + k_m, x_m)v_m)| \leq 1,$$

we conclude that

$$|F(k,x)v)| \leq 1, \quad k \in \mathbb{Z}.$$

Condition B implies that $v = 0$. The contradiction with (2.36) completes the proof. □

Lemma 2.2.9 *There exists a number $\mu > 0$ such that if $(x, v) \in V$ and $|v| \leq \mu$, then $(x, v) \in A$.*

Proof Assuming the contrary, we can find a sequence $(x_m, v_m) \in V$ such that $|v_m| \to 0,\ m \to \infty$, and $(x_m, v_m) \notin A$.

Then

$$\mu_m = \max_{k \geq 0} |F(k, x_m)v_m| > 1$$

(we take into account that $|F(k, x_m)v_m| \to 0,\ k \to \infty$).

Find numbers $k_m > 0$ such that

$$|F(k_m, x_m)v_m| = \mu_m.$$

Since

$$|F(k, x_m)(v_m/\mu_m)| \leq 1, \quad k \geq 0,$$

$(x_m, v_m/\mu_m) \in A$.

The mapping ρ is continuous and $F(k, x)0 = 0$; hence,

$$\max_{0 \leq k \leq K} |F(k, x_m)(v_m/\mu_m)| \to 0, \quad m \to \infty,$$

for any fixed K (note that $x_m \in M$, M is compact, $|v_m| \to 0$, and $\mu_m > 1$).

Hence, $k_m \to \infty$, $m \to \infty$. Lemma 2.2.8 implies now that the relations

$$(x_m, v_m/\mu_m) \in A \quad \text{and} \quad |F(k_m, x_m)(v_m/\mu_m)| = 1$$

are contradictory. □

Lemma 2.2.10 *There exists a number $K > 0$ such that if $(x, v) \in V$, then*

$$|F(k, x)v| \leq (1/2)|v|, \quad k \geq K. \tag{2.37}$$

Proof Apply Lemma 2.2.8 to find a number K such that

$$|F(k, x)v'| < \mu/2, \quad k \geq K,$$

for any $(x, v') \in A$ (where μ is the number from Lemma 2.2.9).

Take any $(x, v) \in V$. If $v \neq 0$, set $v' = \mu(v/|v|)$. Then $(x, v') \in A$ by Lemma 2.2.9, and it follows from Lemma 2.2.8 that

$$|F(k, x)v'| = (\mu/|v|)\,|F(k, x)v| \leq \mu/2, \quad k \geq K,$$

which obviously implies the desired relation (2.37). If $v = 0$, we have nothing to prove. □

Lemma 2.2.11

(1) The subbundles V and W are closed.
(2) There exist numbers $C > 0$ and $\lambda \in (0, 1)$ such that if $(x, v) \in V$, then

$$|F(k, x)v| \leq C\lambda^k |v|, \quad k \geq 0; \tag{2.38}$$

if $(x, v) \in W$, *then*

$$|F(k, x)v| \le C\lambda^{-k}|v|, \quad k \le 0. \tag{2.39}$$

Proof We prove the statements for the subbundle V; for W, the proofs are similar.

To prove statement (1), consider a sequence $(x_k, v_k) \in V$ such that $(x_k, v_k) \to (x, v)$ as $k \to \infty$.

If $v = 0$, then, obviously, $(x, v) \in V$. Assume that $v \ne 0$; then $v_k \ne 0$ for large k, and, by Lemma 2.2.9 there exists a $\mu > 0$ such that

$$(x_k, \mu v_k/|v_k|) \in A.$$

Since A is closed (see Lemma 2.2.7),

$$(x, \mu v/|v|) \in A,$$

and $(x, v) \in V$ by Lemma 2.2.7. This proves the first statement of our lemma.

To prove the second one, apply Lemma 2.2.10 and find a number K such that

$$|F(k, x)v| \le (1/2)|v|, \quad k \ge K, \tag{2.40}$$

for any $(x, v) \in V$.

It follows from (2.40) and from the ρ-invariance of V that

$$|F(2K, x)v| \le (1/2)^2|v|, \dots, |F(kK, x)v| \le (1/2)^k|v|, \quad k \ge 0. \tag{2.41}$$

There exists a number $C_0 > 0$ such that

$$\max_{0 \le k < K,\, x \in M} \|F(k, x)\| \le C_0. \tag{2.42}$$

Let us show that inequality (2.38) holds with $C = 2C_0$ and $\lambda = 2^{1/K}$. We can represent any $k \ge 0$ in the form $k = k_0K + k_1$, where $k_0 \ge 0$ and $0 \le k_1 < K$. If $(x, v) \in V$, then it follows from (2.41) and (2.42) that

$$|F(k, x)v| = |F(k_1, \phi^{k_0K}(x))F(k_0K, x)v| \le C_0(1/2)^{k_0}|v|,$$

but since $k_0 + 1 > k/K$, $-k_0 < -k/K + 1$, and $2^{-k_0} < 2\lambda^k$, we conclude that

$$|F(k, x)v| \le C\lambda^k|v|,$$

as required. □

Remark 2.2.3 Inequalities (2.38) and (2.39) have the same form as inequalities (HSD2.3) and (HSD2.4) in the definition of a hyperbolic set. Thus, if we want to show that some compact, ρ-invariant subset M_0 of M is a hyperbolic set of ρ with subbundles V and W, we only have to show that

$$V_x + W_x = T_xM, \quad x \in M_0. \tag{2.43}$$

Lemma 2.2.12 *Assume that for a sequence $(x_m, v_m) \in TM$ there exists a sequence $k_m \to \infty$ as $m \to \infty$ and a number $r > 0$ such that*

$$|v_m| \le r \quad \text{and} \quad |F(k_m, x_m)v_m| \le r.$$

Then there exists a number $R > 0$ such that

$$|F(k, x_m)v_m| \le R, \quad 0 \le k \le k_m.$$

Proof Assume the contrary, and let there exist $(x_m, v_m) \in TM$ and $k_m \to \infty$ such that

$$b_m := \max_{0 \le k \le k_m} |F(k, x_m)v_m| \to \infty, \quad m \to \infty.$$

Find numbers $l_m \in [0, k_m]$ such that $b_m = |F(l_m, x_m)v_m|$. Since ρ is continuous, it is obvious that

$$l_m \to \infty \quad \text{and} \quad k_m - l_m \to \infty, \quad m \to \infty. \tag{2.44}$$

Set

$$w_m = F(l_m, x_m)(v_m/b_m).$$

Let (x, v) be a limit point of the sequence $(\phi^{l_m}(x_m), w_m)$; then $|v| = 1$. The inequality

$$\left| F(k, \phi^{l_m}(x_m))w_m \right| \le 1$$

holds for $k \in [-l_m, 0] \cup [0, k_m - l_m]$. We apply relations (2.44) and Lemma 2.2.6 (and its analog for W) to conclude that $v \in V_x \cap W_x$, but then $v = 0$ by Condition B. □

Remark 2.2.4 A similar statement is valid if $k_m \to -\infty$. In this case,

$$|F(k, x_m)v_m| \le R, \quad k_m \le k \le 0.$$

Lemma 2.2.13 *If x is a nonwandering point of the diffeomorphism f, then equality (2.43) holds.*

Proof By the definition of a nonwandering point, there exist sequences of points $x_m \in M$ and numbers k_m such that

$$x_m \to x, \quad f^{k_m}(x_m) \to x, \quad |k_m| \to \infty$$

as $m \to \infty$. We may assume that $k_m \to -\infty$.

Consider the linear subspace W_x and let Q be its orthogonal complement. Let $\dim Q = s$. Fix an orthonormal base $v_1, \dots, v_s$ in Q. Clearly, we can find s orthonormal vectors $v_1^m, \dots, v_s^m$ in $T_{x_m}M$ such that $v_j^m \to v_j$ as $m \to \infty$ for $j = 1, \dots, s$.

Let Q_m be the subspace of $T_{x_m}M$ spanned by $v_1^m, \dots, v_s^m$. Introduce the numbers

$$\mu_m = \min\{|F(k_m, x_m)v| : v \in Q_m, |v| = 1\}.$$

We claim that

$$\mu_m \to \infty, \quad m \to \infty. \tag{2.45}$$

If we assume the contrary, we can find a number $r > 0$ and sequences $w_m \in Q_m$, $|w_m| = 1$, and $k_m \to -\infty$ such that

$$|F(k_m, x_m)w_m| \le r.$$

By the remark to Lemma 2.2.12, there exists a number R such that

$$|F(k, x_m)w_m| \le R, \quad k \in [k_m, 0].$$

By Lemma 2.2.6, in this case, any limit point (x, v) of the sequence (x_m, w_m) belongs to W, i.e., $v \in W_x$. This relation contradicts our construction since $w_m \in Q_m$, which implies that v is orthogonal to Q (note that $|v| = 1$). This proves (2.45).

Consider the linear space

$$K_m = F(k_m, x_m)Q_m.$$

Clearly, $K_m \subset T_{y_m}M$, where $y_m = f^{k_m}(x_m)$, and $\dim K_m = s$.

Consider a vector $w \in K_m$, $|w| = 1$. Let $w = F(k_m, x_m)v$. It follows from the definition of the numbers μ_m that

$$|v| \le \mu_m |w| = \mu_m. \tag{2.46}$$

Inequalities (2.46), relations (2.45), and Lemma 2.2.12 imply that for any sequence (y_m, w_m), where $w_m \in K_m$ and $|w_m| = 1$, there exists a number R such that

$$|F(k, x_m)w_m| \le R, \quad k \in [0, -k_m].$$

Now Lemma 2.2.6 implies that any limit point (x, w) of such a sequence (y_m, w_m) belongs to V, i.e., $w \in V_x$.

Select an orthonormal basis $w_1^m, \dots, w_s^m$ in K_m. We may assume that all the sequences $w_1^m, \dots, w_s^m$ converge for some sequence of indices. For definiteness, let

$$w_1^m \to w_1, \dots, w_s^m \to w_s, \quad m \to \infty.$$

The vectors $w_1, \dots, w_s$ are pairwise orthogonal unit vectors in V_x; hence,

$$\dim V_x \geq s. \tag{2.47}$$

By the definition of the spaces Q and Q_m,

$$\dim W_x = n - s.$$

Combining this with inequality (2.47), we see that

$$\dim V_x + \dim W_x \geq n.$$

Since $V_x \cap W_x = \{0\}$ by Condition B, we conclude that

$$V_x + W_x = T_x M,$$

as claimed. □

The nonwandering set of the diffeomorphism f coincides with the nonwandering set of the diffeomorphism $\phi = f^{-1}$. Combining Lemma 2.2.1 with Lemma 2.2.4 applied to the mapping ρ, we conclude that the following statement holds.

Theorem 2.2.2 *If a diffeomorphism f satisfies the analytic strong transversality condition, then the nonwandering set of f is hyperbolic.*

Now we show that the analytic strong transversality condition implies the second part of Axiom A, the density of periodic points in the nonwandering set $\Omega(f)$ of the diffeomorphism f.

Since we are going to use the Mañé theorem in the proof of the implication (the analytic strong transversality condition) $\Rightarrow$ (structural stability) for a diffeomorphism f having the Lipschitz shadowing property, we can essentially simplify this proof (compared to the original Mañé proof) assuming that f has the shadowing property.

Thus, now we prove the following statement.

Theorem 2.2.3 *If a diffeomorphism f has the shadowing property and the nonwandering set $\Omega(f)$ of f is hyperbolic, then periodic points are dense in $\Omega(f)$.*

In this proof, we apply the following two well-known results (see, for example, [71] for their proofs).

First we recall a known definition.

Definition 2.2.2 A homeomorphism f of a metric space (M, dist) is called *expansive* on a set A with *expansivity constant* $a > 0$ if the relations

$$f^k(x), f^k(y) \in A, \quad k \in \mathbb{Z},$$

and

$$\text{dist}\left(f^k(x), f^k(y)\right) \leq a, \quad k \in \mathbb{Z},$$

imply that $x = y$.

Theorem 2.2.4 *If Λ is a hyperbolic set of a diffeomorphism f, then there exists a neighborhood of Λ on which f is expansive.*

Denote by card$A$ the cardinality of a finite or countable set A.

Theorem 2.2.5 (The Birkhoff Constant Theorem) *If the phase space X of a homeomorphism f is compact and U is a neighborhood of the nonwandering set $\Omega(f)$ of f, then there exists a constant $T = T(U)$ such that for any point $x \in X$, the inequality*

$$\textit{card}\left\{k \in \mathbb{Z} : f^k(x) \notin U\right\} \leq T$$

holds.

Proof (of Theorem 2.2.3) Fix an arbitrary point $z \in \Omega(f)$. There exist sequences of points z_n and numbers $l_n \to \infty$ such that

$$z_n \to z \quad \text{and} \quad f^{l_n}(z_n) \to z, \quad n \to \infty.$$

Let U be a neighborhood of the set $\Omega(f)$ on which f is expansive and let a be the corresponding expansivity constant.

Fix an $\varepsilon > 0$ such that the 3ε-neighborhood of $\Omega(f)$ is a subset of U. Denote by U' the 2ε-neighborhood of $\Omega(f)$. We assume, in addition, that $2\varepsilon < a$.

For this ε there exists a $d > 0$ such that any d-pseudotrajectory of f is ε-shadowed by an exact trajectory.

Fix an index n such that

$$\text{dist}(z, z_n), \ \text{dist}(z, f^{l_n}(z_n)) < d/2.$$

Construct a sequence $\{x_k\}$ as follows. Represent $k \in \mathbb{Z}$ in the form $k = k_0 + k_1 l_n$, where $k_1 \in \mathbb{Z}$ and $0 \leq k_0 < l_n$, and set $x_k = f^{k_0}(z_n)$.

Clearly, the sequence $\{x_k\}$ is periodic with period l_n; the choice of n implies that this sequence is a d-pseudotrajectory of f.

We claim that

$$\{x_k\} \subset U'. \tag{2.48}$$

Assuming the contrary, we can find an index m such that $x_m \notin U'$, i.e.,

$$\operatorname{dist}(x_m, \Omega(f)) \geq 2\varepsilon,$$

but then

$$\operatorname{dist}(x_{m+kl_n}, \Omega(f)) \geq 2\varepsilon, \quad k \in \mathbb{Z}. \tag{2.49}$$

Let $p \in M$ be a point whose trajectory ε-shadows $\{x_k\}$, i.e.,

$$\operatorname{dist}\left(f^k(p), x_k\right) < \varepsilon, \quad k \in \mathbb{Z};$$

let $p_k = f^k(p)$.

Then it follows from inequalities (2.49) that

$$\operatorname{dist}(p_{m+kl_n}, \Omega(f)) \geq \varepsilon, \quad k \in \mathbb{Z},$$

which contradicts Theorem 2.2.5. Thus, we have established inclusion (2.48).

Set $r = f^{l_n}(p)$. Since $x_k = x_{k+l_n}$, the following inequalities hold:

$$\operatorname{dist}\left(f^k(r), x_k\right) = \operatorname{dist}\left(f^{k+l_n}(p), x_{k+l_n}\right) < \varepsilon, \quad k \in \mathbb{Z}.$$

Then

$$\operatorname{dist}\left(f^k(r), f^k(p)\right) < 2\varepsilon < a, \quad k \in \mathbb{Z};$$

in addition, inclusion (2.48) implies that

$$f^k(r), f^k(p) \in U, \quad k \in \mathbb{Z}.$$

Since f is expansive on U, $r = p$.

Thus, p is a periodic point of f.

Since ε and d can be taken arbitrarily small, there is such a point p in an arbitrarily small neighborhood of the point z. □

Thus, it remains to show that the analytic strong transversality condition implies the strong transversality condition (stable and unstable manifolds of nonwandering points are transverse).

For this purpose, we apply the following well-known theorem on the behavior of trajectories of a diffeomorphism in a neighborhood of a hyperbolic set (its proof can be easily reduced to Theorem 6.4.9 in the book [28]).

Theorem 2.2.6 *Let Λ be a hyperbolic set of a diffeomorphism f with hyperbolicity constants C, λ. For any $C_1 > C$ and $\lambda_1 \in (\lambda, 1)$ there exists a neighborhood U of Λ with the following property. If $x \in W^s(p)$, $p \in \Lambda$, and $f^k(x) \in U$ for $k \geq 0$,* *then there exist two complementary linear subspaces $L^+(x)$ and $L^-(x)$ of T_xM such that*

(1)

$$L^+(x) = T_x W^s(p),\ L^-(x) = T_x W^u(p);$$

(2)

$$\left|Df^k(x)v\right| \le C_1 \lambda_1^k |v|, \quad k \ge 0,\ v \in L^+(x),$$

and

$$\left|Df^k(x)v\right| \ge (1/C_1)\lambda_1^{-k}|v|, \quad k \ge 0,\ v \in L^-(x).$$

Remark 2.2.5 Of course, a similar statement holds if $x \in W^u(p)$, $p \in \Lambda$, and $f^k(x)$ belongs to a small neighborhood of Λ for $k \le 0$.

Clearly, it is enough for us to prove that if $r \in W^s(p) \cap W^u(q)$, where $p, q \in \Omega(f)$, then

$$B^+(r) \subset T_r W^s(p) \text{ and } B^-(r) \subset T_r W^u(q). \tag{2.50}$$

We prove the first inclusion in (2.50) by proving that

$$B^+(r) \subset L^+(r) \tag{2.51}$$

and applying Theorem 2.2.6; the second inclusion is proved in a similar way.

Any trajectory of a diffeomorphism satisfying Axiom A tends to one of the basic sets as time tends to $\pm\infty$ (see Theorem 1.3.2).

Take as Λ the basic set to which $f^k(r)$ tends as $k \to \infty$; obviously, p belongs to this basic set. Of course, we may assume that the positive semitrajectory of r belongs to a neighborhood of Λ having the properties described in Theorem 2.2.6.

Assume that inclusion (2.51) does not hold; take $v \in B^+(r) \setminus L^+(r)$ and represent

$$v = v^s + v^u, \quad v^s \in L^+(r),\ v^u \in L^-(r);$$

then $v^u \ne 0$.

Then

$$\left|Df^k v\right| \ge \left|Df^k v_u\right| - \left|Df^k v_s\right| \ge (1/C_1)\lambda^{-k}|v^u| - C_1\lambda^k|v^s| \to \infty, \quad k \to \infty,$$

which contradicts the relation defining $B^+(r)$.

We have completely proved the Mañé theorem.

Historical Remarks In his paper [39], R. Mañé gave several equivalent characterizations of structural stability of a diffeomorphism; Theorem 1.3.7 of this book is just one of them.

The property of expansivity of a dynamical system with discrete time is now one of the classical properties studied in the global theory of dynamical systems. Theorem 2.2.4 is folklore. Let us mention J. Ombach's paper [49] in which it was shown (see Proposition 9) that a compact invariant set Λ of a diffeomorphism f is hyperbolic if and only if $f|_\Lambda$ is expansive and has the (standard) shadowing property (compare with Sect. 4.1).

Theorem 2.2.5 was proved in G. Birkhoff's book [10].

2.3 Diffeomorphisms with Lipschitz Shadowing

Our main result in this section is as follows.

Theorem 2.3.1 *If a diffeomorphism of class C^1 of a smooth closed n-dimensional manifold M has the Lipschitz shadowing property, then f is structurally stable.*

As stated in Theorem 1.4.1 (1), a structurally stable diffeomorphism f has the Lipschitz shadowing property. Combining this statement with Theorem 2.3.1, we conclude that for diffeomorphisms, structural stability is equivalent to Lipschitz shadowing.

Proof (of Theorem 2.3.1) Let us first explain the main idea of the proof.

Fix an arbitrary point $p \in M$, consider its trajectory $\{p_k = f^k(p) : k \in \mathbb{Z}\}$, and denote $A_k = Df(p_k)$. Consider the sequence $\mathscr{A} = \{A_k : k \in \mathbb{Z}\}$.

In Sect. 2.1 devoted to the Maizel' and Pliss theorems, we worked with sequences $\mathscr{A}$ of isomorphisms of Euclidean spaces. Here we apply these theorems (and all the corresponding notions of the Perron property etc.) to the sequences $\mathscr{A} = \{Df(p_k)\}$ (see the remark concluding Sect. 3.1).

We claim that if f has the Lipschitz shadowing property, then $\mathscr{A}$ has the Perron property on $\mathbb{Z}$.

By the Maizel' theorem, the Perron property on $\mathbb{Z}$ implies that the sequence $\mathscr{A}$ is hyperbolic on both "rays" $\mathbb{Z}_-$ and $\mathbb{Z}_+$. Denote by $S_k^-, U_k^-, k \in \mathbb{Z}_-$ and $S_k^+, U_k^+, k \in \mathbb{Z}_+$ the corresponding stable and unstable subspaces.

Then, by the Pliss theorem, the subspaces U_0^- and S_0^+ are transverse.

Clearly,

$$|A_k \circ \cdots \circ A_0 v| \to 0, \quad v \in S_0^+, k \to \infty,$$

and

$$\left|(A_k)^{-1} \circ \cdots \circ (A_0)^{-1} v\right| \to 0, \quad v \in U_0^-, k \to -\infty,$$

which means that $U_0^- \subset B^-(p)$ and $S_0^+ \subset B^+(p)$, where $B^-(p)$ and $B^+(p)$ are the subspaces from the analytic transversality condition.

The transversality of the subspaces U_0^- and S_0^+ implies the transversality of the subspaces $B^-(x)$ and $B^+(x)$. Since x is arbitrary, f is structurally stable by the Mañé theorem.

Now we prove our claim.

To clarify the reasoning, we first prove an analog of this result, Lemma 2.3.2, for a diffeomorphism of the Euclidean space $\mathbb{R}^n$. Of course, $\mathbb{R}^n$ is not compact, but we avoid the appearing difficulty making the following additional assumption (and noting that an analog of this assumption is certainly valid for a diffeomorphism of class C^1 of a closed smooth manifold). We call the condition below *Condition S*.

Thus, we assume that for any $\mu > 0$ we can find a $\delta = \delta(\mu) > 0$ (independent of k) such that if $|v| \le \delta$, then

$$|f(p_k + v) - A_k v - p_{k+1}| \le \mu |v|, \quad k \in \mathbb{Z}. \tag{2.52}$$

The basic technical part of the proof of Lemma 2.3.2 is the following statement (Lemma 2.3.1). In the following two Lemmas, 2.3.1 and 2.3.2, f is a diffeomorphism of $\mathbb{R}^n$ that has the Lipschitz shadowing property with constants $\mathcal{L}, d_0 > 0$, $\{p_k = f^k(p)\}$ is an arbitrary trajectory of f, $A_k = Df(p_k)$, and it is assumed that Condition S is satisfied.

Lemma 2.3.1 *Fix a natural number N. For any sequence*

$$w_k \in \mathbb{R}^n, \quad k \in \mathbb{Z},$$

with $|w_k| < 1$ there exists a sequence

$$z_k \in \mathbb{R}^n, \quad k \in \mathbb{Z},$$

such that

$$|z_k| \le \mathcal{L} + 1, \quad k \in \mathbb{Z}, \tag{2.53}$$

and

$$z_{k+1} = A_k z_k + w_{k+1}, \quad -N \le k \le N. \tag{2.54}$$

Proof Thus, we assume that f has the Lipschitz shadowing property with constants $\mathcal{L}, d_0 > 0$.

Define vectors

$$\Delta_k \in \mathbb{R}^n, \quad -N \le k \le N + 1,$$

by the following relations:

$$\Delta_{-N} = 0 \quad \text{and} \quad \Delta_{k+1} = A_k \Delta_k + w_{k+1}, \quad -N \le k \le N. \tag{2.55}$$

Clearly, there exists a number Q (depending on N, $\mathscr{A}$, and w_k) such that

$$|\Delta_k| \le Q, \quad -N \le k \le N+1. \tag{2.56}$$

Fix a small number $d \in (0, d_0)$ (we will reduce this number during the proof) and consider the following sequence $\xi = \{x_k \in \mathbb{R}^n : k \in \mathbb{Z}\}$:

$$x_k = \begin{cases} f^{k+N}(p_{-N}), & k < -N; \\ p_k + d\Delta_k, & -N \le k \le N+1; \\ f^{k-N-1}(p_{N+1} + d\Delta_{N+1}), & k > N+1. \end{cases}$$

Note that if $-N \le k \le N$, then

$$|x_{k+1} - f(x_k)| = |p_{k+1} + d\Delta_{k+1} - f(p_k + d\Delta_k)| \le$$

$$\le d\,|\Delta_{k+1} - A_k\Delta_k| + |f(p_k + d\Delta_k) - p_{k+1} - dA_k\Delta_k|.$$

Since we consider a finite number of w_k, the condition $|w_k| < 1$ implies that there is a $\mu \in (0, 1)$ such that the first term above does not exceed μd; by Condition S, the second term is less than $(1 - \mu)d$ if d is small. Hence, in this case, the sum is less than d.

For the remaining values of k,

$$|x_{k+1} - f(x_k)| = 0.$$

Thus, we may take $d \le d_0$ so small that ξ is a d-pseudotrajectory of f. Then there exists a trajectory $\eta = \{y_k : k \in \mathbb{Z}\}$ of f such that

$$|x_k - y_k| \le \mathscr{L}d, \quad k \in \mathbb{Z}. \tag{2.57}$$

Denote $t_k = (y_k - p_k)/d$. Since $\Delta_k = (x_k - p_k)/d$, it follows from (2.57) that

$$|\Delta_k - t_k| = |x_k - y_k|/d \le \mathscr{L}, \quad k \in \mathbb{Z}. \tag{2.58}$$

It follows from (2.56) and (2.57) that

$$|y_k - p_k| \le |y_k - x_k| + |x_k - p_k| \le (\mathscr{L} + Q)d, \quad k \in \mathbb{Z}.$$

Hence,

$$|t_k| \le \mathscr{L} + Q, \quad k \in \mathbb{Z}. \tag{2.59}$$

Now we define a finite sequence

$$b_k \in \mathbb{R}^n, \quad -N \le k \le N+1,$$

by the following relations:

$$b_{-N} = t_{-N} \quad \text{and} \quad b_{k+1} = A_k b_k, \quad -N \le k \le N. \tag{2.60}$$

Take $\mu_1 \in (0,1)$ such that

$$\left((K+1)^{2N} + (K+1)^{2N-1} + \cdots + 1\right)\mu_1 < 1, \tag{2.61}$$

where $K = \sup \|A_k\|$. Set

$$\mu = \frac{\mu_1}{\mathscr{L} + Q}$$

and consider d so small that inequality (2.52) holds for $|v| \le \delta$ with $\delta = (\mathscr{L} + Q)d$.

The definition of the vectors t_k implies that

$$dt_{k+1} = y_{k+1} - p_{k+1} = f(y_k) - f(p_k) = f(p_k + dt_k) - f(p_k).$$

Since $|dt_k| \le (\mathscr{L} + Q)d$ by (2.59), it follows from Condition S and from the above choice of d that

$$|dt_{k+1} - dA_k t_k| = |f(p_k + dt_k) - f(p_k) - dA_k t_k| \le$$

$$\le \mu |dt_k| \le \mu(\mathscr{L} + Q)d = \mu_1 d.$$

Hence,

$$t_{k+1} = A_k t_k + \theta_k, \quad \text{where} \quad |\theta_k| < \mu_1. \tag{2.62}$$

Consider the vectors

$$c_k = t_k - b_k.$$

Note that $c_{-N} = 0$ by (2.60) and

$$c_{k+1} = A_k c_k + \theta_k, \quad \text{where} \quad |\theta_k| < \mu_1$$

by (2.62).

Thus,

$$|c_{-N+1}| \leq |\theta_{-N}| < \mu_1,$$

$$|c_{-N+2}| \leq |A_{-N+1}c_{-N+1} + \theta_{-N+1}| \leq (K+1)\mu_1,$$

and so on, which implies the estimate

$$|c_k| \leq \left((K+1)^{2N} + (K+1)^{2N-1} + \cdots + 1\right)\mu_1 < 1, \quad -N \leq k \leq N.$$

Hence,

$$|t_k - b_k| \leq 1, \quad -N \leq k \leq N. \tag{2.63}$$

Finally, we consider the sequence

$$z_k = \begin{cases} 0, & k < -N; \\ \Delta_k - b_k, & -N \leq k \leq N+1; \\ 0, & k > N+1. \end{cases}$$

Relations (2.55) and (2.60) imply relations (2.54); estimates (2.58) and (2.63) imply estimate (2.53). □

Lemma 2.3.2 *The sequence $\mathscr{A} = \{A_k\}$ has the Perron property.*

Proof Take an arbitrary sequence

$$w_k \in \mathbb{R}^n, \quad k \in \mathbb{Z},$$

with $|w_k| < 1$ and prove that an analog of Eq. (2.54) has a solution

$$z_k \in \mathbb{R}^n, \quad k \in \mathbb{Z},$$

with

$$|z_k| \leq \mathscr{L} + 1, \quad k \in \mathbb{Z}.$$

Fix a natural N and consider the sequence

$$w_k^{(N)} = \begin{cases} w_k, & -N \leq k \leq N; \\ 0, & |k| \geq N+1. \end{cases}$$

By Lemma 2.3.1, there exists a sequence $\left\{z_k^{(N)}, \ k \in \mathbb{Z}\right\}$ such that

$$z_{k+1}^{(N)} = A_k z_k^{(N)} + w_k^{(N)}, \quad -N \leq k \leq N, \tag{2.64}$$

and

$$\left|z_k^{(N)}\right| \leq \mathscr{L} + 1, \quad k \in \mathbb{Z}. \tag{2.65}$$

Passing to a subsequence of $\left\{z_k^{(N)}\right\}$, we can find a sequence $\{v_k\}$ such that

$$v_k = \lim_{N\to\infty} z_k^{(N)}, \quad k \in \mathbb{Z}.$$

(Note that do not assume uniform convergence.) Passing to the limit in (2.64) and (2.65) as $N \to \infty$, we see that

$$v_{k+1} = A_k y_k + w_k, \quad k \in \mathbb{Z},$$

and

$$|v_k| \leq \mathscr{L} + 1, \quad k \in \mathbb{Z}.$$

Thus, we have shown that the sequence $\mathscr{A}$ has the Perron property. □

Now let us explain how to prove the required statement in the case of a smooth closed manifold M.

Lemma 2.3.3 *If a diffeomorphism of class C^1 of a smooth closed n-dimensional manifold M has the Lipschitz shadowing property, $\{p_k = f^k(p)\}$ is an arbitrary trajectory of f, and $A_k = Df(p_k)$, then the sequence $\mathscr{A} = \{A_k\}$ has the Perron property.*

Proof Let exp be the standard exponential mapping on the tangent bundle of M generated by the fixed Riemannian metric dist. Let

$$\exp_x : T_xM \to M$$

be the corresponding exponential mapping at a point $x \in M$.

Denote (just for this proof) by $B(r, x)$ the ball in M of radius r centered at a point x; let $B_T(r, x)$ be the ball in T_xM of radius r centered at the origin.

It is well known that there exists an $r > 0$ such that for any $x \in M$, $\exp_x$ is a diffeomorphism of $B_T(r, x)$ onto its image and $\exp_x^{-1}$ is a diffeomorphism of $B(r, x)$ onto its image; in addition, $D\exp_x(0) = \text{Id}$.

Thus, we may assume that r is chosen so that the following inequalities hold for any $x \in M$:

$$\text{dist}(\exp_x(v), \exp_x(w)) \leq 2|v - w|, \quad v, w \in B_T(r, x), \tag{2.66}$$

and

$$\left|\exp_x^{-1}(y) - \exp_x^{-1}(z)\right| \leq 2\text{dist}(y, z), \quad y, z \in B(r, x). \tag{2.67}$$

These inequalities mean that distances are distorted not more than twice when we pass from the manifold to its tangent space or from the tangent space to the manifold (if we work in a small neighborhood of a point of the manifold or in a small neighborhood of the origin of the tangent space).

In our reasoning below, we always assume that d is so small that the corresponding points belong to such small neighborhoods.

Now we fix a trajectory $\{p_k = f^k(p)\}$ of our diffeomorphism f and introduce the mappings

$$F_k = \exp_{p_{k+1}}^{-1} \circ f \circ \exp_{p_k} : T_{p_k}M \to T_{p_{k+1}}M.$$

Clearly,

$$DF_k(0) = A_k.$$

The analog of Condition S is as follows: For any $\mu > 0$ we can find a $\delta > 0$ (independent of k) such that if $|v| < \delta$, then

$$|F_k(v) - A_k v| \le \mu|v|, \quad k \in \mathbb{Z}. \tag{2.68}$$

Of course, this condition is satisfied automatically since f is of class C^1 and the manifold M is compact.

To prove that the sequence $\mathscr{A}$ has the Perron property, let us consider the difference equations

$$v_{k+1} = A_k y_k + w_k, \quad k \in \mathbb{Z}, \tag{2.69}$$

where $v_k \in T_{p_k}M$ and $w_k \in T_{p_{k+1}}M$.

We assume that $|w_k| < 1$, $k \in \mathbb{Z}$. Let us "translate" the reasoning of Lemma 2.3.1 to the "manifold language."

We fix a natural N and consider the sequence

$$\Delta_k \in T_{p_k}M, \quad -N \le k \le N+1,$$

defined by relations (2.55). Let Q satisfy (2.56).

We fix a small d and define the sequence $\xi = \{x_k \in M : k \in \mathbb{Z}\}$ by

$$x_k = \begin{cases} f^{k+N}(p_{-N}), & k < -N; \\ \exp_{p_k}(d\Delta_k), & -N \le k \le N+1; \\ f^{k-N-1}(\exp_{p_{N+1}}(d\Delta_{N+1})), & k > N+1. \end{cases}$$

This definition and inequalities (2.66) imply that if d is small enough, then

$$\operatorname{dist}\left(x_{k+1}, \exp_{p_{k+1}}(dA_k\Delta_k)\right) < 2d.$$

Since

$$f(x_k) = \exp_{p_{k+1}}(F_k(d\Delta_k)),$$

condition (2.68) with $\mu < 1$ implies that

$$\text{dist}\left(\exp_{p_{k+1}}(dA_k\Delta_k), f(x_k)\right) < 2d,$$

and we see that

$$\text{dist}\,(f(x_k), x_{k+1}) < 4d.$$

Thus, there exists an exact trajectory $\eta = \{y_k : k \in \mathbb{Z}\}$ of f such that

$$\text{dist}(x_k, y_k) \leq 4\mathcal{L}d, \quad k \in \mathbb{Z}. \tag{2.70}$$

Now we consider the finite sequence

$$t_k = \frac{1}{d}\exp_{p_k}^{-1}(y_k), \quad -N \leq k \leq N.$$

Inequalities (2.70) and (2.67) imply that

$$|\Delta_k - t_k| \leq 8\mathcal{L}, \quad k \in \mathbb{Z}. \tag{2.71}$$

Note that

$$\text{dist}(y_k, p_k) \leq \text{dist}(y_k, x_k) + \text{dist}(x_k, p_k) \leq (4\mathcal{L} + 2Q)d, \quad k \in \mathbb{Z}.$$

Hence,

$$|t_k| \leq 8\mathcal{L} + 4Q, \quad k \in \mathbb{Z}.$$

Now we define a finite sequence

$$b_k \in T_{p_k}M, \quad -N \leq k \leq N+1,$$

by relations (2.60) and repeat the reasoning of Lemma 2.3.1 with

$$\mu = \frac{\mu_1}{8\mathcal{L} + 4Q},$$

where μ_1 is the same as above (see relation (2.61)).

The rest of the proof is literally the same (with natural replacement of $\mathbb{R}^n$ by the corresponding tangent spaces), and we get the relation

$$|t_k - b_k| < 1$$

similar to (2.63).

Finally, we get the estimate

$$|z_k| \leq 8\mathcal{L} + 1,$$

which completes the proof of the analog of Lemma 2.3.1.

The rest of the proof of the implication "Lipschitz shadowing property implies the Perron property of the sequence $\mathscr{A}$" almost literally repeats the proof of Lemma 2.3.2. □

Historical Remarks Theorem 2.3.1 was published by the first author and S. B. Tikhomirov in the paper [68]. Let us mention that the paper [67] contained the first proof of the fact that structural stability follows from certain shadowing property based on a combination of the Maizel', Pliss, and Mañé theorems.

2.4 Lipschitz Periodic Shadowing for Diffeomorphisms

The main result of this section is as follows.

Theorem 2.4.1 *A diffeomorphism f of class C^1 of a smooth closed n-dimensional manifold M has the Lipschitz periodic shadowing property if and only if f is Ω-stable.*

First we prove the "if" statement of Theorem 2.4.1.

Theorem 2.4.2 *If a diffeomorphism f is Ω-stable, then f has the Lipschitz periodic shadowing property.*

Let us give one more definition.

Definition 2.4.1 We say that a diffeomorphism f has the *Lipschitz shadowing property* on a set U if there exist positive constants $\mathcal{L}, d_0$ such that if $\xi = \{x_i : i \in \mathbb{Z}\} \subset U$ is a d-pseudotrajectory with $d \leq d_0$, then there exists a point $p \in U$ such that inequalities (1.5) hold.

Remark 2.4.1 It follows from Theorems 1.4.2 and 2.2.4 that we can find a neighborhood U of a hyperbolic set Λ of a diffeomorphism f having the above-formulated property and such that f is expansive on U.

We start by proving several auxiliary results.

Lemma 2.4.1 *Let f be a homeomorpism of a compact metric space $(M, dist)$. For any neighborhood U of the nonwandering set $\Omega(f)$ there exist positive numbers*

T, δ_1 such that if $\xi = \{x_i : i \in \mathbb{Z}\}$ is a d-pseudotrajectory of f with $d \le \delta_1$ and

$$x_k, x_{k+1}, \dots, x_{k+l} \notin U$$

for some $k \in \mathbb{Z}$ and $l > 0$, then $l \le T$.

Proof Take a neighborhood U of the nonwandering set $\Omega(f)$ and let T be the Birkhoff constant for the homeomorphism f given for this neighborhood by Theorem 2.2.5. Assume that there does not exist a number δ_1 with the desired property; then there exists a sequence $d_j \to 0$ as $j \to \infty$ and a sequence of d_j-pseudotrajectories $\{x_k^{(j)} : k \in \mathbb{Z}\}$ of f such that

$$\left\{x_k^{(j)} : 0 \le k \le T-1\right\} \cap U = \emptyset$$

for all j.

The set $M' = M \setminus U$ is compact. Passing to a subsequence, if necessary, we may assume that $x_0^{(j)} \to x_0$ as $j \to \infty$. In this case,

$$x_k^{(j)} \to f^k(x_0) \in M', \quad 0 \le k \le T-1,$$

and we get a contradiction with the choice of T. □

Now let us recall some basic properties of Ω-stable diffeomorphisms. It was noted in Sect. 1.3 that a diffeomorphism f is Ω-stable if and only if f satisfies Axiom A and the no cycle condition (Theorem 1.3.3).

Let $\Omega_1, \dots, \Omega_m$ be the basic sets in decomposition (1.15) of the nonwandering set of an Ω-stable diffeomorphism f.

Below we need one folklore technical statement. Recall that we write $\Omega_i \to \Omega_j$ if there is a point $x \notin \Omega(f)$ such that

$$f^{-k}(x) \to \Omega_i \text{ and } f^k(x) \to \Omega_j, \quad k \to \infty.$$

Theorem 2.4.3 *Assume that a diffeomorphism f is Ω-stable. For any family of neighborhoods U_i of the basic sets Ω_i one can find neighborhoods $V_i \subset U_i$ such that if a point x belongs to some V_i and there exist indices $0 < l \le m$ such that*

$$f^l(x) \notin U_i \text{ and } f^m(x) \in V_j,$$

then there exist basic sets $\Omega_{i_1}, \dots, \Omega_{i_t}$ such that

$$\Omega_i \to \Omega_{i_1} \to \dots \to \Omega_{i_t} \to \Omega_j. \tag{2.72}$$

Proof Reducing the given neighborhoods U_i, we may assume that the compact sets $U_i' = f(\mathrm{Cl}(U_i)) \cup \mathrm{Cl}(U_i)$ are disjoint.

Assume that our statement does not hold. In this case, there exist sequences of points x_k, $k \geq 0$, and indices $l(k) \leq m(k)$ such that

$$x_k \to \Omega_i, \quad f^{l(k)}(x_k) \notin U_i, \quad f^{m(k)}(x_k) \to \Omega_i, \quad k \to \infty.$$

Clearly, we may assume that

$$x_k, f(x_k), \dots, f^{l(k)-1}(x_k) \in U_i$$

while

$$y_k := f^{l(k)}(x_k) \notin U_i.$$

Then $y_k \in U_i'$, and, passing to a subsequence, if necessary, we may assume that $y_k \to y \in U_i'$ as $k \to \infty$.

Since Ω_i is a compact f-invariant set, $l(k) \to \infty$ as $k \to \infty$. Thus, for any $t < 0$, $f^t(y_k) \in U_i$ for large k, and it follows that $f^t(y) \in \mathrm{Cl}(U_i)$ for any $t < 0$. We note that the set $\mathrm{Cl}(U_i)$ intersects a single basic set, Ω_i, and refer to (1.16) to conclude that

$$y \in W^u(\Omega_i). \tag{2.73}$$

By the same relation (1.16), there exists a basic set Ω_{i_1} such that

$$y \in W^s(\Omega_{i_1}). \tag{2.74}$$

By our choice of U_i, the sets $\mathrm{Cl}(f(U_i)) \setminus U_i$ do not contain nonwandering points. Thus, if $i_1 = i$, inclusions (2.73) and (2.74) mean the existence of a 1-cycle, and we get the desired contradiction.

Hence, $i_1 \neq i$ and $\Omega_i \to \Omega_{i_1}$. Consider the compact set

$$Y = \{f^k(y) : k \geq 0\} \cup \Omega_{i_1}.$$

Clearly, the set Y has a neighborhood Z such that $U_{i_1} \subset Z$ and Z does not intersect a small neighborhood of Ω_i.

Since $y_k = f^{l_k}(x_k) \to y$, there exist indices $l_1(k)$ such that

$$f^t(y_k) = f^{l(k)+t}(x_k) \in Z, \quad 0 \leq t \leq l_1(k),$$

for large k, and

$$x_{1,k} = f^{l_1(k)}(y_k) = f^{l(k)+l_1(k)}(x_k) \to \Omega_{i_1}, \quad k \to \infty.$$

At the same time, the positive trajectories of the points y_k (and hence, of the points $x_{1,k}$) must leave Z (and hence, U_{i_1}) since the sequence

$$f^{m(k)-l(k)}(y_k) = f^{m(k)}(x_k)$$

tends to Ω_i.

Thus, we can repeat the above reasoning with the points $x_{1,k}$ and the basic set Ω_{i_1} instead of x_k and Ω_i.

Such a process will produce basic sets $\Omega_{i_1}, \Omega_{i_2}, \dots$ such that

$$\Omega_i \to \Omega_{i_1} \to \Omega_{i_2} \to \dots.$$

Since f has no cycles, this process is finite, and, as a result, we conclude that there exist basic sets $\Omega_{i_1}, \dots, \Omega_{i_t}$ such that relations (2.72) hold. □

Now we apply the above theorem to prove a statement concerning periodic pseudotrajectories of Ω-stable diffeomorphisms.

Lemma 2.4.2 *Assume that a diffeomorphism f is Ω-stable. For any family of disjoint neighborhoods W_i of the basic sets Ω_i there exists a number $\delta_2 > 0$ such that any periodic d-pseudotrjectory ξ of f with $d \le \delta_2$ belongs to a single neighborhood W_i.*

Proof Fix arbitrary disjoint neighborhoods W_i of the basic sets Ω_i and find a number $\varepsilon > 0$ and neighborhoods U_i of Ω_i such that

$$N(\varepsilon, U_i) \subset W_i, \quad i = 1, \dots, m.$$

Apply Theorem 2.4.3 to find for U_i the corresponding neighborhoods V_i of Ω_i. Reducing ε, if necessary, we can find neighborhoods V_i' of Ω_i such that

$$N(\varepsilon, V_i') \subset V_i, \quad i = 1, \dots, m.$$

By Lemma 2.4.1, there exist positive numbers T, δ_1 such that if $\xi = \{x_k\}$ is a d-pseudotrajectory of f with $d \le \delta_1$ and

$$x_k, x_{k+1}, \dots, x_{k+l} \notin V := \bigcup_{i=1}^{m} V_i'$$

for some $k \in \mathbb{Z}$ and $l > 0$, then $l \le T$.

Find a number $\delta_2 \in (0, \delta_1)$ such that if $\xi = \{x_k\}$ is a d-pseudotrajectory of f with $d \le \delta_2$, then

$$\operatorname{dist}(f^l(x_k), x_{k+l}) < \varepsilon, \quad 0 \le l \le T + 1,$$

for any $k \in \mathbb{Z}$.

Now let $\xi = \{x_k\}$ be a periodic d-pseudotrajectory of f of period μ with $d \leq \delta_2$. Let us call a V-block of ξ a finite segment

$$\xi_{k,m} = \{x_k, x_{k+1}, \dots, x_{k+m}\}, \quad k \in \mathbb{Z}, \ m > 0,$$

such that $x_k, x_{k+m} \in V$ while $x_{k+l} \notin V$ for $0 < l < m$. Note that in this case, $m \leq T + 1$.

Let us note simple properties of V-blocks.

It follows from the choice of δ_2 that if $\xi_{k,m}$ is a V-block for which there exist indices $i, j \in \{1, \dots, m\}$ such that $x_k \in V_i'$ and $x_{k+m} \in V_j'$, then $\operatorname{dist}(f^m(x_k), x_{k+m}) < \varepsilon$; hence, $f^m(x_k) \in V_j$.

At the same time, if for such a V-block there exists an index $l \in (0, m)$ such that $x_{k+l} \notin W_i$, then $\operatorname{dist}(f^l(x_k), x_{k+l}) < \varepsilon$; hence, $f^l(x_k) \notin U_i$.

It follows from Theorem 2.4.3 that in this case, there exists a relation of the form (2.72); the absence of cycles implies that $j \neq i$.

Since $\delta_2 < \delta_1$, there exists a neighborhood V_i' such that ξ intersects V_i'.

Changing indices of ξ, we may assume that $x_0 \in V_i'$.

If either $x_k \in W_i$ for $k \geq 0$ or any V-block $\xi_{k,m}$ with $k \geq 0$ belongs to W_i, then the statement of our lemma follows from the periodicity of ξ.

It was noted above that if $\xi_{k,m}$ be a V-block with $x_k \in V_j$ for $k \geq 0$ for which there exists an index $l \in (0, m)$ such that $x_{k_l} \notin W_j$, then there exists an index $j' \neq j$ for which we have a relation

$$\Omega_j \to \dots \to \Omega_{j'}$$

of the form (2.72).

Thus, if we assume that there exists a V-block $\xi_{k,m}$ with $k \geq 0$ such that $\xi_{k,m} \setminus W_i \neq \emptyset$, then we get an index $j_1 \neq i$ such that we have a relation

$$\Omega_i \to \dots \to \Omega_{j_1}$$

of the form (2.72).

Going to "the right" of this V-block $\xi_{k,m}$ and continuing this process, we construct a sequence of pairs of indices $(i, j_1), (j_1, j_2), \dots$ such that

$$\Omega_i \to \dots \to \Omega_{j_1}, \quad \Omega_{j_1} \to \dots \to \Omega_{j_2}, \quad \dots.$$

In this case, it follows from the absence of cycles that all the indices $i, j_1, j_2, \dots$ are different.

But the μ-periodicity of ξ implies that if $\xi_{k,m}$ is a V-block and n is a natural number, then $\xi_{k+n\mu,m}$ is an identical V-block, and the existence of the above sequence with different $i, j_1, j_2, \dots$ is impossible.

Now we prove Theorem 2.4.2.

By Remark 2.4.1, there exist disjoint neighborhoods $U_1, \dots, U_m$ of the basic sets $\Omega_1, \dots, \Omega_m$ such that

(i) f has the Lipschitz shadowing property on any of U_j with the same constants $\mathscr{L}, d_0^*$;
(ii) f is expansive on any of U_j with the same expansivity constant a.

Find neighborhoods W_j of Ω_j (and reduce d_0^*, if necessary) so that the $\mathscr{L}d_0^*$-neighborhoods of W_j belong to U_j. Apply Lemma 2.4.2 to find the corresponding constant δ_2.

We claim that f has the Lipschitz periodic shadowing property with constants $\mathscr{L}, d_0$, where

$$d_0 = \min\left(d_0^*, \delta_2, \frac{a}{2\mathscr{L}}\right).$$

Take a μ-periodic d-pseudotrajectory $\xi = \{x_k\}$ of f with $d \leq d_0$. Lemma 2.4.2 implies that there exists a neighborhood W_i such that $\xi \subset W_i \subset U_i$.

Thus, there exists a point p such that inequalities (1.5) hold. Let us show that p is a periodic point of f. By the choice of U_i and $W_i, f^k(p) \in U_i$ for all $k \in \mathbb{Z}$. Let $q = f^\mu(p)$. Inequalities (1.5) and the periodicity of ξ imply that

$$\operatorname{dist}\left(f^k(q), x_k\right) = \operatorname{dist}\left(f^{k+\mu}(p), x_k\right) = \operatorname{dist}\left(f^{k+\mu}(p), x_{k+\mu}\right) \leq \mathscr{L}d, \quad k \in \mathbb{Z}.$$

Thus,

$$\operatorname{dist}\left(f^k(q), f^k(p)\right) \leq 2\mathscr{L}d \leq a, \quad k \in \mathbb{Z},$$

which implies that $f^\mu(p) = q = p$. This completes the proof. □

Now we prove the "only if" statement of Theorem 2.4.1.

Theorem 2.4.4 *If a diffeomorphism f has the Lipschitz periodic shadowing property, then f is Ω-stable.*

Thus, let us assume that f has the Lipschitz periodic shadowing property (with constants $\mathscr{L} \geq 1, d_0 > 0$). Clearly, in this case f^{-1} has the Lipschitz periodic shadowing property as well (and we assume that the constants $\mathscr{L}, d_0$ are the same for f and f^{-1}).

To clarify the presentation, in the construction of pseudotrajectories in the following Lemmas 2.4.3 and 2.4.4, we assume that f is a diffeomorphism of $\mathbb{R}^n$ (and leave to the reader consideration of the case of a manifold).

We also assume that there exists a number $N > 0$ such that $\|Df(x)\| \leq N$ for all considered points x (an analog of this assumption is satisfied in the case of a closed manifold).

Recall that we denote by $\mathrm{Per}(f)$ the set of periodic points of f.

Lemma 2.4.3 *Every point $p \in Per(f)$ is hyperbolic.*

Proof To get a contradiction, let us assume that f has a nonhyperbolic periodic point p (to simplify notation, we assume that p is a fixed point; literally the same reasoning can be applied to a periodic point of period $m > 1$). In addition, we assume that $p = 0$.

In this case, we can represent

$$f(v) = Av + F(v),$$

where $A = Df(0)$ and $F(v) = o(v)$ as $v \to 0$.

By our assumption, A is a nonhyperbolic matrix. The following two cases are possible:

Case 1: A has a real eigenvalue λ with $|\lambda| = 1$;
Case 2: A has a complex eigenvalue λ with $|\lambda| = 1$.

We treat in detail only Case 1 and give a comment concerning Case 2. To simplify presentation, we assume that 1 is an eigenvalue of A; the case of eigenvalue -1 is treated similarly.

We can introduce coordinate v such that, with respect to this coordinate, the matrix A has block-diagonal form,

$$A = \mathrm{diag}(B, P), \tag{2.75}$$

where B is a Jordan block of size $l \times l$:

$$B = \begin{pmatrix} 1\ 1\ 0 \dots 0 \\ 0\ 1\ 1 \dots 0 \\ \vdots\ \vdots\ \vdots\ \ddots\ \vdots \\ 0\ 0\ 0 \dots 1 \end{pmatrix}.$$

Of course, introducing new coordinates, we have to change the constants $\mathcal{L}$ and d_0; we denote the new constants by the same symbols. In addition, we assume that $\mathcal{L}$ is integer.

We start considering the case $l = 2$; in this case,

$$B = \begin{pmatrix} 1\ 1 \\ 0\ 1 \end{pmatrix}.$$

Let

$$e_1 = (1, 0, 0, \dots, 0) \text{ and } e_2 = (0, 1, 0, \dots, 0)$$

be the first two vectors of the standard orthonormal basis.

Let $K = 7\mathcal{L}$.

Take a small $d > 0$ and construct a finite sequence $y_0, \dots, y_Q$ of points (where Q is determined later) as follows: $y_0 = 0$ and

$$y_{k+1} = Ay_k + de_2, \quad k = 0, \dots, K-1. \tag{2.76}$$

Then

$$y_K = (Z_1(K)d, Kd, 0, \dots, 0),$$

where the natural number $Z_1(K)$ is determined by K (we do not write $Z_1(K)$ explicitly). Now we set

$$y_{k+1} = Ay_k - de_2, \quad k = K, \dots, 2K-1.$$

Then

$$y_{2K} = (Z_2(K)d, 0, 0, \dots, 0),$$

where the natural number $Z_2(K)$ is determined by K as well. Take $Q = 2K + Z_2(K)$; if we set

$$y_{k+1} = Ay_k - de_1, \quad k = 2K, \dots, Q-1,$$

then $y_Q = 0$. Let us note that both numbers Q and

$$Y := \frac{\max_{0 \le k \le Q-1} |y_k|}{d}$$

are determined by K (and hence, by $\mathscr{L}$).

Now we construct a Q-periodic sequence $x_k, k \in \mathbb{Z}$, that coincides with the above sequence for $k = 0, \dots, Q$.

We claim that if d is small enough, then $\xi = \{x_k\}$ is a $2d$-pseudotrajectory of f (and this pseudotrajectory is Q-periodic by construction).

Indeed, we know that $|x_k| \le Yd$ for $k \in \mathbb{Z}$. Since $F(v) = o(|v|)$ as $|v| \to 0$,

$$|F(x_k)| < d, \quad k \in \mathbb{Z}, \tag{2.77}$$

if d is small enough.

The definition of x_k implies that

$$|x_{k+1} - Ax_k| = d, \quad k \in \mathbb{Z}. \tag{2.78}$$

It follows from (2.77) and (2.78) that

$$|x_{k+1} - f(x_k)| \le |x_{k+1} - Ax_k| + |F(x_k)| < 2d,$$

which implies that $\xi = \{x_k\}$ is a $2d$-pseudotrajectory of f if d is small enough.

Now we estimate the distances between points of trajectories of the diffeomorphism f and its linearization at zero.

Let us take a vector p_0 and assume that the sequence $p_k = f^k(p_0)$ belongs to the ball $|v| \leq (Y + 2\mathscr{L})d$ for $0 \leq k \leq K$. Let $r_k = A^k p_0$ (we impose no conditions on r_k since below we estimate F at points q_k only).

Take a small number $\mu \in (0, 1)$ (to be chosen later) and assume that d is small enough, so that the inequality

$$|F(v)| \leq \mu |v|$$

holds for $|v| \leq (Y + 2\mathscr{L})d$.

By our assumption, $\|A\| = \|Df(0)\| \leq N$. Then

$$|p_1| \leq |Ap_0| + |F(p_0)| \leq (N + 1)|p_0|, \ldots,$$

$$|p_k| \leq |Ap_{k-1}| + |F(p_{k-1})| \leq (N + 1)^k |p_0|$$

for $1 \leq k \leq K$, and

$$|p_1 - r_1| = |Ap_0 + F(p_0) - Ap_0| \leq \mu |p_0|,$$

$$|p_2 - r_2| = |Ap_1 + F(p_1) - Ar_1| \leq N|p_1 - r_1| + \mu |p_1| \leq \mu(2N + 1)|p_0|,$$

$$|p_3 - r_3| \leq N|p_2 - r_2| + \mu |p_2| \leq \mu(N(2N + 1) + (N + 1)^2)|p_0|,$$

and so on.

Thus, there exists a number $\nu = \nu(K, N)$ such that

$$|p_k - r_k| \leq \mu\nu |p_0|, \quad 0 \leq k \leq K.$$

We take $\mu = 1/\nu$, note that $\mu = \mu(K, N)$, and get the inequalities

$$|p_k - r_k| \leq |p_0|, \quad 0 \leq k \leq K, \tag{2.79}$$

for d small enough.

Since f has the Lipschitz periodic shadowing property, for d small enough, the Q-periodic $2d$-pseudotrajectory ξ is $2\mathscr{L}d$-shadowed by a periodic trajectory. Let p_0 be a point of this trajectory such that

$$|p_k - x_k| \leq \mathscr{L}d, \quad k \in \mathbb{Z}, \tag{2.80}$$

where $p_k = f^k(p_0)$.

The inequalities $|x_k| \leq Yd$ and (2.80) imply that

$$|p_k| \leq |x_k| + |p_k - x_k| \leq (Y + 2\mathscr{L})d, \quad k \in \mathbb{Z}. \tag{2.81}$$

Note that $|p_0| \leq 2\mathscr{L}d$.

Set $r_k = A^k p_0$; we deduce from estimate (2.79) that if d is small enough, then

$$|p_K - r_K| \leq |p_0| \leq 2\mathscr{L}d. \tag{2.82}$$

Denote by $v^{(2)}$ the second coordinate of a vector v.

It follows from the structure of the matrix A that

$$\left|r_K^{(2)}\right| = \left|p_0^{(2)}\right| \leq 2\mathscr{L}d. \tag{2.83}$$

The relations

$$\left|y_K^{(2)}\right| = Kd \text{ and } |p_K - y_K| \leq 2\mathscr{L}d$$

imply that

$$\left|p_K^{(2)}\right| \geq Kd - 2\mathscr{L}d = 5\mathscr{L}d \tag{2.84}$$

(recall that $K = 7\mathscr{L}$).

Estimates (2.82)–(2.84) are contradictory. Our lemma is proved in Case 1 for $l = 2$.

If $l = 1$, then the proof is simpler; the first coordinate of $A^k v$ equals the first coordinate of v, and we construct the periodic pseudotrajectory perturbing the first coordinate only.

If $l > 2$, the reasoning is parallel to that above; we first perturb the lth coordinate to make it Kd, and then produce a periodic sequence consequently making zero the lth coordinate, the $(l-1)$st coordinate, and so on.

If λ is a complex eigenvalue, $\lambda = a + bi$, we take a real 2×2 matrix

$$R = \begin{pmatrix} a & -b \\ b & a \end{pmatrix}$$

and assume that in representation (2.75), B is a real $2l \times 2l$ Jordan block:

$$B = \begin{pmatrix} R & E_2 & 0 & \dots & 0 \\ 0 & R & E_2 & \dots & 0 \\ \vdots & \vdots & \vdots & \ddots & \vdots \\ 0 & 0 & 0 & \dots & R \end{pmatrix},$$

where E_2 is the 2×2 identity matrix.

After that, almost the same reasoning works; we note that $|Rv| = |v|$ for any 2-dimensional vector v and construct periodic pseudotrajectories replacing, for example, formulas (2.76) by the formulas

$$y_{k+1} = Ay_k + dw_k, \quad k = 0, \dots, K-1,$$

where jth coordinates of the vector w_k are zero for $j = 1, \dots, 2l-2, 2l+1, \dots, n$, while the 2-dimensional vector corresponding to $(2l-1)$st and $2l$th coordinates has the form $R^k w$ with $|w| = 1$, and so on. We leave details to the reader. The lemma is proved. □

Lemma 2.4.4 *There exist constants $C > 0$ and $\lambda \in (0, 1)$ depending only on N and $\mathscr{L}$ and such that, for any point $p \in Per(f)$, there exist complementary subspaces $S(p)$ and $U(p)$ of $\mathbb{R}^n$ that are Df-invariant, i.e.,*

(H1) $Df(p)S(p) = S(f(p))$ and $Df(p)U(p) = U(f(p))$,
and the inequalities
(H2.1) $|Df^j(p)v| \le C\lambda^j |v|, \quad v \in S(p), j \ge 0$,
and
(H2.2) $|Df^{-j}(p)v| \le C\lambda^j |v|, \quad v \in U(p), j \ge 0$,
hold.

Remark 2.4.2 This lemma means that the set Per(f) has all the standard properties of a hyperbolic set, with the exception of compactness.

Proof Take a periodic point $p \in$ Per(f); let m be the minimal period of p.

Denote $p_i = f^i(p)$, $A_i = Df(p_i)$, and $B = Df^m(p)$. It follows from Lemma 2.4.3 that the matrix B is hyperbolic. Denote by $S(p)$ and $U(p)$ the invariant subspaces of B corresponding to parts of its spectrum inside and outside the unit disk, respectively. Clearly, $S(p)$ and $U(p)$ are invariant with respect to Df, they are complementary subspaces of $\mathbb{R}^n$, and the following relations hold:

$$\lim_{n\to+\infty} B^n v_s = \lim_{n\to+\infty} B^{-n} v_u = 0, \quad v_s \in S(p), v_u \in U(p). \tag{2.85}$$

We prove that inequalities (H2.2) hold with $C = 4\mathscr{L}$ and $\lambda = 1 + 1/(2\mathscr{L})$ (inequalities (H2.1) are established by similar reasoning applied to f^{-1} instead of f).

Consider an arbitrary nonzero vector $v_u \in U(p)$ and an integer $j \ge 0$. Define sequences of vectors v_i, e_i and numbers $\lambda_i > 0$ for $i \ge 0$ as follows:

$$v_0 = v_u, \quad v_{i+1} = A_i v_i, \quad e_i = \frac{v_i}{|v_i|}, \quad \lambda_i = \frac{|v_{i+1}|}{|v_i|} = |A_i e_i|.$$

Let

$$\tau = \frac{\lambda_{m-1} \cdot \ldots \cdot \lambda_1 + \lambda_{m-1} \cdot \ldots \cdot \lambda_2 + \ldots + \lambda_{m-1} + 1}{\lambda_{m-1} \cdot \ldots \cdot \lambda_0}.$$

Consider the sequence $\{a_i \in \mathbb{R} : i \geq 0\}$ defined by the following formulas:

$$a_0 = \tau, \quad a_{i+1} = \lambda_i a_i - 1. \tag{2.86}$$

Note that

$$a_m = 0 \quad \text{and} \quad a_i > 0, \quad i \in [0, m-1]. \tag{2.87}$$

Indeed, if $a_i \leq 0$ for some $i \in [0, m-1]$, then $a_k < 0$ for $k \in [i+1, m]$.

It follows from (2.85) that there exists an $n > 0$ such that

$$|B^{-n}\tau e_0| < 1. \tag{2.88}$$

Consider the finite sequence of vectors $\{w_i : \ i \in [0, m(n+1)]\}$ defined as follows:

$$\begin{cases} w_i = a_i e_i, & i \in [0, m-1]; \\ w_m = B^{-n}\tau e_0; \\ w_{m+1+i} = A_i w_{m+i}, & i \in [0, mn-1]. \end{cases}$$

Clearly,

$$w_{km} = B^{k-1-n}\tau e_0, \quad k \in [1, n+1],$$

which means that we can consider $\{w_i\}$ as an $m(n+1)$-periodic sequence defined for $i \in \mathbb{Z}$.

Let us note that

$$A_i w_i = a_i A_i e_i = a_i \frac{v_{i+1}}{|v_i|}, \quad i \in [0, m-2],$$

$$w_{i+1} = (\lambda_i a_i - 1)\frac{v_{i+1}}{|v_{i+1}|} = a_i \frac{v_{i+1}}{|v_i|} - e_{i+1}, \quad i \in [0, m-2],$$

and

$$A_{m-1}w_{m-1} = a_{m-1}\frac{v_m}{|v_{m-1}|} = \frac{v_m}{\lambda_{m-1}|v_{m-1}|} = e_m$$

(in the last relation, we take into account that $a_{m-1}\lambda_{m-1} = 1$ since $a_m = 0$).

The above relations and condition (2.88) imply that

$$|w_{i+1} - A_i w_i| < 2, \quad i \in \mathbb{Z}. \tag{2.89}$$

Now we take a small $d > 0$ and consider the $m(n+1)$-periodic sequence

$$\xi = \{x_i = p_i + dw_i : i \in \mathbb{Z}\}.$$

We claim that if d is small enough, then ξ is a $2d$-pseudotrajectory of f.

Represent

$$f(x_i) = f(p_i) + Df(p_i)dw_i + F_i(dw_i) = p_{i+1} + A_i dw_i + F_i(dw_i),$$

where $F_i(v) = o(|v|)$ as $v \to 0$.

It follows from estimates (2.77) that

$$|f(x_i) - x_{i+1}| < 2d$$

for small d.

By Lemma 2.4.3, the m-periodic trajectory $\{p_i\}$ is hyperbolic; hence, $\{p_i\}$ has a neighborhood in which $\{p_i\}$ is the unique periodic trajectory. It follows that if d is small enough, then the pseudotrajectory $\{x_i\}$ is $2\mathscr{L}d$-shadowed by $\{p_i\}$.

The inequalities $|x_i - p_i| \le 2\mathscr{L}d$ imply that $|a_i| = |w_i| \le 2\mathscr{L}$ for $0 \le i \le m-1$. Now the equalities $\lambda_i = (a_{i+1}+1)/a_i$ imply that if $0 \le i \le m-1$, then

$$\lambda_0 \cdot \ldots \cdot \lambda_{i-1} = \frac{a_1+1}{a_0}\frac{a_2+1}{a_1} \cdots \frac{a_i+1}{a_{i-1}} =$$

$$= \frac{a_i+1}{a_0}\left(1+\frac{1}{a_1}\right)\cdots\left(1+\frac{1}{a_{i-1}}\right) \ge$$

$$\ge \frac{1}{2\mathscr{L}}\left(1+\frac{1}{2\mathscr{L}}\right)^{i-1} > \frac{1}{4\mathscr{L}}\left(1+\frac{1}{2\mathscr{L}}\right)^{i}$$

(we take into account that $1 + 1/(2\mathscr{L}) < 2$ since $\mathscr{L} \ge 1$).

It remains to note that

$$|Df^i(p)v_u| = \lambda_{i-1} \cdots \lambda_0 |v_u|, \quad 0 \le i \le m-1,$$

and that we started with an arbitrary vector $v_u \in U(p)$.

This proves our statement for $j \le m-1$. If $j \ge m$, we take an integer $k > 0$ such that $km > j$ and repeat the above reasoning for the periodic trajectory $p_0, \ldots, p_{km-1}$ (note that we have not used the condition that m is the minimal period). The lemma is proved. □

In the following lemmas, we return to the case of a diffeomorphism f of a smooth closed manifold M since the reasoning becomes "global." We still assume that f has the Lipschitz periodic shadowing property and apply analogs of Lemmas 2.4.3 and 2.4.4 for the case of a manifold.

Lemma 2.4.5 *The diffeomorphism f satisfies Axiom A.*

Proof Denote by P_l the set of points $p \in \mathrm{Per}(f)$ of index l (as usual, the index of a hyperbolic periodic point is the dimension of its stable manifold).

Let R_l be the closure of P_l. Clearly, R_l is a compact f-invariant set. We claim that any R_l is a hyperbolic set. Let $n = \dim M$.

Consider a point $q \in R_l$ and fix a sequence of points $p_m \in P_l$ such that $p_m \to q$ as $m \to \infty$. By an analog of Lemma 2.4.4, there exist complementary subspaces $S(p_m)$ and $U(p_m)$ of $T_{p_m}M$ (of dimensions l and $n - l$, respectively) for which estimates (H2.1) and (H2.2) hold.

Standard reasoning shows that, introducing local coordinates in a neighborhood of (q, T_qM) in the tangent bundle of M, we can select a subsequence p_{m_k} for which the sequences $S(p_{m_k})$ and $U(p_{m_k})$ converge (in the Grassmann topology) to subspaces of T_qM (let S_0 and U_0 be the corresponding limit subspaces).

The limit subspaces S_0 and U_0 are complementary in T_qM. Indeed, consider the "angle" β_{m_k} between the subspaces $S(p_{m_k})$ and $U(p_{m_k})$ which is defined (with respect to the introduced local coordinates in a neighborhood of (q, T_qM)) as follows:

$$\beta_{m_k} = \min |v^s - v^u|,$$

where the minimum is taken over all possible pairs of unit vectors $v^s \in S(p_{m_k})$ and $v^u \in U(p_{m_k})$.

The same reasoning as in the proof of Lemma 2.1.5 shows that the values β_{m_k} are estimated from below by a positive constant $\alpha = \alpha(N, C, \lambda)$. Clearly, this implies that the subspaces S_0 and U_0 are complementary.

It is easy to show that the limit subspaces S_0 and U_0 are unique (which means, of course, that the sequences $S(p_m)$ and $U(p_m)$ converge). For the convenience of the reader, we prove this statement.

To get a contradiction, assume that there is a subsequence p_{m_i} for which the sequences $S(p_{m_i})$ and $U(p_{m_i})$ converge to complementary subspaces S_1 and U_1 different from S_0 and U_0 (for definiteness, we assume that $S_0 \setminus S_1 \neq \emptyset$).

Due to the continuity of Df, the inequalities

$$\left|Df^j(q)v\right| \leq C\lambda^j |v|, \quad v \in S_0 \cup S_1,$$

and

$$\left|Df^j(q)v\right| \geq C^{-1}\lambda^{-j} |v|, \quad v \in U_0 \cup U_1,$$

hold for $j \geq 0$.

Since

$$T_qM = S_0 \oplus U_0 = S_1 \oplus U_1,$$

our assumption implies that there is a vector $v \in S_0$ such that

$$v = v^s + v^u, \quad v^s \in S_1, v^u \in U_1, v^u \neq 0.$$

Then

$$|Df^j(q)v| \leq C\lambda^j|v| \to 0, \quad j \to \infty,$$

and

$$\left|Df^j(q)v\right| \geq C^{-1}\lambda^{-j}|v^u| - C\lambda^j|v^s| \to \infty, \quad j \to \infty,$$

and we get the desired contradiction.

It follows that there are uniquely defined complementary subspaces $S(q)$ and $U(q)$ for $q \in R_l$ with proper hyperbolicity estimates; the Df-invariance of these subspaces is obvious. We have shown that each R_l is a hyperbolic set with $\dim S(q) = l$ and $\dim U(q) = n - l$ for $q \in R_l$.

If $r \in \Omega(f)$, then there exists a sequence of points $r_m \to r$ as $m \to \infty$ and a sequence of indices $k_m \to \infty$ as $m \to \infty$ such that $f^{k_m}(r_m) \to r$.

Clearly, if we continue the sequence

$$r_m, f(r_m), \dots, f^{k_m-1}(r_m)$$

periodically with period k_m, we get a periodic d_m-pseudotrajectory of f with $d_m \to 0$ as $m \to \infty$.

Since f has the Lipschitz periodic shadowing property, for large m there exist periodic points p_m such that $\text{dist}(p_m, r_m) \to 0$ as $m \to \infty$. Thus, periodic points are dense in $\Omega(f)$.

Since hyperbolic sets with different dimensions of the subspaces $U(q)$ are disjoint, we get the equality

$$\Omega(f) = R_0 \cup \dots \cup R_n,$$

which implies that $\Omega(f)$ is hyperbolic. The lemma is proved. □

Thus, to prove Theorem 2.4.4, it remains to prove the following lemma.

Lemma 2.4.6 *If f has the Lipschitz periodic shadowing property, then f satisfies the no cycle condition.*

Proof To simplify presentation, we prove that f has no 1-cycles (in the general case, the idea is literally the same, but the notation is heavy; we leave this case to the reader).

To get a contradiction, assume that

$$p \in (W^u(\Omega_i) \cap W^s(\Omega_i)) \setminus \Omega(f).$$

In this case, there are sequences of indices $j_m, k_m \to \infty$ as $m \to \infty$ such that

$$f^{-j_m}(p), f^{k_m}(p) \to \Omega_i, \quad m \to \infty.$$

Since the set Ω_i is compact, we may assume that

$$f^{-j_m}(p) \to q \in \Omega_i \text{ and } f^{k_m}(p) \to r \in \Omega_i.$$

Since Ω_i contains a dense positive semitrajectory, there exist points $s_m \to r$ and indices $l_m > 0$ such that $f^{l_m}(s_m) \to q$ as $m \to \infty$.

Clearly, if we continue the sequence

$$p, f(p), \dots, f^{k_m-1}(p), s_m, \dots, f^{l_m-1}(s_m), f^{-j_m}(p), \dots, f^{-1}(p)$$

periodically with period $k_m + l_m + j_m$, we get a periodic d_m-pseudotrajectory of f with $d_m \to 0$ as $m \to \infty$.

Since f has the Lipschitz periodic shadowing property, there exist periodic points p_m (for m large enough) such that $p_m \to p$ as $m \to \infty$, and we get the desired contradiction with the assumption that $p \notin \Omega(f)$. The lemma is proved. □

Historical Remarks Theorem 2.4.1 was published by A. V. Osipov, the first author, and S. B. Tikhomirov in [50].

2.5 Hölder Shadowing for Diffeomorphisms

In this section, we explain the main ideas of the proof of the following result.

Theorem 2.5.1 *Assume that a diffeomorphism f of class C^2 of a smooth closed manifold has the Hölder shadowing property on finite intervals with constants $\mathscr{L}, C, d_0, \theta, \omega$ and that*

$$\theta \in (1/2, 1) \text{ and } \theta + \omega > 1. \tag{2.90}$$

Then f is structurally stable.

The proof of Theorem 2.5.1 is quite complicated. For that reason, we try to simplify the presentation and omit inessential technical details; the reader can find the original Tikhomirov's proof in the paper [101].

The main two steps of the proof of Theorem 2.5.1 are as follows.

First one considers a trajectory $\{p_k = f^k(p)\}$ of f, denotes $A_k = Df(p_k)$, and shows that under conditions of Theorem 2.5.1, the sequence $\mathscr{A} = \{A_k\}$ has a weak analog of the Perron property (in which the existence of bounded solutions of the inhomogeneous difference equations is replaced by the existence of "slowly growing" solutions).

We reproduce this part of the proof in Theorem 2.5.2 in which we restrict our consideration to the case of a diffeomorphism f of the Euclidean space $\mathbb{R}^n$.

After that, it is shown that the above-mentioned weak analog of the Perron property implies then f satisfies the analytic strong transversality condition (with exponential estimates) and, hence, by the Mañé theorem, f is structurally stable. To explain the basic techniques of that part of the proof, we prove the above statement in Theorem 2.5.3 in the case of a one-dimensional phase space (and note that the reasoning in the proof of Theorem 2.5.3 reproduces the most important part of the proof given by Tikhomirov). We again refer the reader to [101] for the proof of the general case.

Theorem 2.5.2 *Assume that a diffeomorphism f of the Euclidean space $\mathbb{R}^n$ has the Hölder shadowing property on finite intervals with constants $\mathscr{L}, C, d_0, \theta, \omega$ and that condition (2.90) is satisfied.*

Assume, in addition, that there exist constants $S, \varepsilon > 0$ such that

$$|f(p_k + v) - p_{k+1} - A_k v| \le S|v|^2, \quad k \in \mathbb{Z}, \ |v| \le \varepsilon. \tag{2.91}$$

Then there exist constants $L > 0$ and $\gamma \in (0, 1)$ such that for any $i \in \mathbb{Z}$ and $N > 0$ and any sequence

$$W = \{w_k \in \mathbb{R}^n : \ i + 1 \le k \le i + N + 1\} \tag{2.92}$$

with $|w_k| \le 1$, the difference equations

$$v_{k+1} = A_k v_k + w_{k+1}, \quad i \le k \le i + N, \tag{2.93}$$

have a solution

$$V = \{v_k : \ i \le k \le i + N + 1\}, \tag{2.94}$$

such that the value

$$\|V\| := \max_{i \le k \le i+N+1} |v_k| \tag{2.95}$$

satisfies the estimate

$$\|V\| \le L N^{\gamma}. \tag{2.96}$$

Remark 2.5.1 Clearly, an analog of condition (2.91) is satisfied if we consider a diffeomorphism of class C^2 for which the trajectory $\{p_k\}$ is contained in a bounded subset of $\mathbb{R}^n$ (or a diffeomorphism of class C^2 of a smooth closed manifold studied in the original paper [101]). In fact, it was noted by Tikhomirov that one can prove a similar result in the case where exponent 2 in (2.91) is replaced by any $\nu > 1$. The reasoning remains almost the same, but calculations become very cumbersome. For that reason, we follow the proof given in [101] (with exponent 2).

Proof (of Theorem 2.5.2) Denote

$$\alpha = \theta - 1/2.$$

Inequalities (2.90) imply that

$$\alpha \in (0, 1/2) \text{ and } 1/2 - \alpha < \omega. \tag{2.97}$$

Consider two auxiliary linear functions of $\beta \geq 0$,

$$g_1(\beta) = (2 + \beta)(1/2 - \alpha) \text{ and } g_1(\beta) = (2 + \beta)\omega.$$

By inequalities (2.97),

$$g_2(0) = 2\omega > 1 - 2\alpha = g_1(0) \in (0, 1)$$

and

$$g_2'(\beta) = \omega > 1/2 - \alpha = g_1'(\beta).$$

Hence, there exists a $\beta > 0$ such that

$$g_1(\beta) \in (0, 1) \text{ and } g_2(\beta) > 1.$$

We fix such a β and write the above relations in the form

$$0 < (2 + \beta)(1/2 - \alpha) < 1 \text{ and } (2 + \beta)\omega > 1. \tag{2.98}$$

Introduce the values

$$\gamma = ((2 + \beta)\omega)^{-1} \text{ and } \gamma_1 = 1 - (2 + \beta)(1/2 - \alpha).$$

Then it follows from (2.98) that

$$0 < \gamma < 1 \text{ and } \gamma_1 > 0. \tag{2.99}$$

Now we fix a sequence W of the form (2.92) and denote by $E(W)$ the set of all sequences V of the form (2.94) that satisfy Eqs. (2.93). The function $\|V\|$ is continuous on the linear space of sequences V; the set $E(W)$ is closed. Hence, the value

$$F(W) = \min_{V \in E(W)} \|V\| \tag{2.100}$$

is defined.

The set of finite sequences W of the form (2.92) with

$$\|W\| = \max_{i+1 \le k \le N+1} |w_k| \le 1$$

is compact. The function $F(W)$ is continuous in W; thus, there exists the number

$$Q = \max_W F(W).$$

Let us fix sequences W_0 and $V_0 \in E(W_0)$ such that

$$Q = F(W_0) = \|V_0\|. \tag{2.101}$$

Note the following two properties of the number Q. They follow from the definition of Q and from the linearity of Eqs. (2.93).

(Q1) Any sequence $V \in E(W_0)$ satisfies the inequality

$$\|V\| \ge Q.$$

(Q2) For any sequence W of the form (2.92) there exists a sequence $V \in E(W)$ such that

$$\|V\| \le Q\|W\|.$$

It follows from property (Q2) that to complete the proof of our theorem, it is enough to prove the following statement:

There exists a number L independent of i and N such that

$$Q \le LN^{\gamma}. \tag{2.102}$$

Define the number

$$d = \varepsilon Q^{-(2+\beta)}. \tag{2.103}$$

Let us consider the following two cases.

Case 1: $C((S+1)d)^{-\omega} < N$. In this case,

$$Q < (\varepsilon^{\omega}(S+1)^{\omega}/C)^{\gamma} N^{\gamma},$$

which proves inequalities (2.103) with

$$L = (\varepsilon^{\omega}(S+1)^{\omega}/C)^{\gamma}.$$

Case 2: $C((S+1)d)^{-\omega} \geq N$. In this case, we prove a stronger statement: There exists a number L independent of i and N such that

$$Q \leq L. \tag{2.104}$$

Treating Case 2, we assume without loss of generality that $i = 0$.

Also, without loss of generality, we assume that $\varepsilon < 1$ and $Q > 2$. Concerning the latter assumption, we note that if there exists a fixed number L independent of N such that $Q \leq L$, then estimate (2.104) is obviously valid. Thus, we may assume that Q is larger than any prescribed number independent of N. Applying the same reasoning, we assume that Q is so large that

$$Q > ((S+1)\varepsilon/d_0)^{1/(2+\beta)} \tag{2.105}$$

and

$$\mathcal{L}((S+1)\varepsilon/Q^{2+\beta})^{\theta} < \varepsilon/2. \tag{2.106}$$

Fix sequences W_0 and V_0 for which relation (2.101) is valid. To simplify notation, write $V_0 = \{v_k\}$.

Consider the sequence of points

$$y_k = p_k + dv_k, \quad 0 \leq k \leq N+1.$$

We claim that this sequence is an $(S+1)d$-pseudotrajectory of f.

Let us first note that $|v_k| \leq Q$; hence,

$$|dv_k| \leq \varepsilon Q^{-(2+\beta)}Q = \varepsilon Q^{-(1+\beta)} < \varepsilon/2. \tag{2.107}$$

In addition,

$$(dQ)^2 = (\varepsilon Q^{-(1+\beta)})^2 < \varepsilon Q^{-(2+\beta)} = d. \tag{2.108}$$

Now we estimate

$$|f(y_k) - y_{k+1}| = |f(p_k + dv_k) - (p_{k+1} + dv_{k+1})| =$$

$$= |f(p_k + dv_k) - (p_{k+1} + dA_k v_k + dw_{k+1})| \leq$$

$$\leq |f(p_k + dv_k) - (p_{k+1} + dA_k v_k)| + d|w_{k+1}| \leq$$

$$\leq S|dv_k|^2 + d \leq (S+1)d.$$

We estimate the first term of the third line taking into account condition (2.91) and inequality (2.107); estimating the first term of the last line, we refer to inequality (2.108).

Inequality (2.105) implies that

$$Q^{2+\beta} > (S+1)\varepsilon/d_0;$$

hence,

$$(S+1)d = (S+1)\varepsilon Q^{-(2+\beta)} < d_0.$$

Since we treat Case 2,

$$N \leq C((S+1)d)^{-\omega} < Cd^{-\omega},$$

and we can apply the Hölder shadowing property on finite intervals to conclude that there exists an exact trajectory $\{x_k\}$ of f such that

$$|y_k - x_k| \leq \mathscr{L}((S+1)d)^{\theta}, \quad 0 \leq k \leq N+1.$$

Denote $x_k = p_k + c_k$ and $\mathscr{L}_1 = \mathscr{L}(S+1)^{\theta}$. Then

$$|dv_k - c_k| \leq |y_k - x_k| \leq \mathscr{L}_1 d^{\theta}, \quad 0 \leq k \leq N+1, \tag{2.109}$$

and

$$|c_k| \leq Qd + \mathscr{L}_1 d^{\theta}, \quad 0 \leq k \leq N+1. \tag{2.110}$$

Inequalities (2.107) and (2.106) imply that

$$|c_k| < \varepsilon.$$

By the first inequality in (2.98),

$$Q > Q^{(1/2-\alpha)(2+\beta)} = (\varepsilon/d)^{1/2-\alpha} = \varepsilon^{1/2-\alpha} d^{\alpha-1/2}.$$

Hence,

$$Qd > \varepsilon^{1/2-\alpha} d^{\alpha+1/2} = \varepsilon^{1/2-\alpha} d^{\theta}.$$

Now it follows from (2.110) that there exists an $\mathscr{L}_2$ independent of N such that

$$|c_k| \le \mathscr{L}_2 Qd.$$

Since $p_{k+1} + c_{k+1} = f(p_k + c_k)$, we can estimate

$$|c_{k+1} - A_k c_k| = |f(p_k + c_k) - (p_{k+1} + A_k c_k| \le S|c_k)|^2 \le S\mathscr{L}_2(Qd)^2.$$

Denote $t_{k+1} = c_{k+1} - A_k c_k$; then

$$|t_k| \le S|c_k|^2 \le \mathscr{L}_3(Qd)^2,$$

where the constant $\mathscr{L}_3$ does not depend on N. By property (Q2), there exists a sequence z_k such that

$$z_{k+1} = A_k z_k + t_{k+1} \text{ and } |z_k| \le Q\mathscr{L}_3(Qd)^2, \quad 0 \le k \le N.$$

Consider the sequence $r_k = c_k - z_k$. Clearly,

$$r_{k+1} = A_k r_k \text{ and } |r_k - c_k| \le Q\mathscr{L}_3(Qd)^2, \quad 0 \le k \le N. \tag{2.111}$$

Now we define the sequence $e_k = (dv_k - r_k)/d$. Relations (2.109) and (2.111) imply that

$$e_{k+1} = A_k e_k + w_{k+1}, \quad 0 \le k \le N, \tag{2.112}$$

and

$$|e_k| = |((dv_k - c_k) - (r_k - c_k))/d| \le \mathscr{L}_1 d^{\theta-1} + \mathscr{L}_3 Q^3 d, \quad 0 \le k \le N.$$

Property (Q1) implies that

$$\mathscr{L}_1 d^{\theta-1} + \mathscr{L}_3 Q^3 d = \mathscr{L}_1 d^{\alpha-1/2} + \mathscr{L}_3 Q^3 d \ge Q.$$

We can apply (2.103) and find $\mathscr{L}_4, \mathscr{L}_5 > 0$ independent of N and such that this inequality takes the form

$$\mathscr{L}_4 Q^{-(2+\beta)(\alpha-1/2)} + \mathscr{L}_5 Q^{1-\beta} \ge Q,$$

or

$$\mathscr{L}_4 Q^{1-\gamma_1} + \mathscr{L}_5 Q^{1-\beta} \ge Q.$$

It follows that either

$$\mathscr{L}_4 Q^{1-\gamma_1} \ge Q/2$$

or

$$\mathcal{L}_5 Q^{1-\beta} \geq Q/2,$$

which implies that

$$Q \leq \max\left((2\mathcal{L}_4)^{1/\gamma_1}, (2\mathcal{L}_5)^{1/\beta}\right).$$

The theorem is proved. □

Now we assume, in addition, that there exists a constant $R > 0$ such that

$$\|A_k\| \leq R, \quad k \in \mathbb{Z}. \tag{2.113}$$

Remark 2.5.2 Of course, an estimate of the form (2.113) holds for $A_k = Df(p_k)$ in the case of a diffeomorphism f of a closed manifold.

Theorem 2.5.3 *Let f be a diffeomorphism of the line $\mathbb{R}$ having the Hölder shadowing property on finite intervals. Assume that conditions (2.91) and (2.113) are satisfied for a trajectory $\{p_k = f^k(p)\}$. There exists a constant $\mu \in (0, 1)$ with the following property.*

For any $k \in \mathbb{Z}$ there exists a constant $C > 0$ and subspaces $S(p_k)$ and $U(p_k)$ of $\mathbb{R}$ such that

$$S(p_k) + U(p_k) = \mathbb{R}, \tag{2.114}$$

$$|A_{k+l-1} \cdots A_k v| \leq C\mu^l |v|, \quad v \in S(p_k),\ l \geq 0, \tag{2.115}$$

$$|A_{k-l}^{-1} \cdots A_{k-1}^{-1} v| \leq C\mu^l |v|, \quad v \in U(p_k),\ l \geq 0. \tag{2.116}$$

The essential part of the proof of Theorem 2.5.3 is contained in the following lemma.

Let us first introduce some notation. Consider a one-dimensional vector (i.e., a real number) e_0 with $|e_0| = 1$ and define a sequence $\{e_k : k \in \mathbb{Z}\}$ as follows:

$$e_{k+1} = A_k e_k / |A_k e_k|, \quad e_{-k-1} = A_{-k-1}^{-1} e_{-k} / |A_{-k-1}^{-1} e_{-k}|, \quad k \geq 0. \tag{2.117}$$

Set

$$\lambda_k = |A_k e_k|.$$

It follows from inequalities (2.113) that

$$\lambda_k \in [1/R, R], \quad k \in \mathbb{Z}. \tag{2.118}$$

Set also

$$\Pi(k, l) = \lambda_k \cdots \lambda_{k+l-1}, \quad k \in \mathbb{Z},\ l \geq 1. \tag{2.119}$$

Lemma 2.5.1 *If the sequence $\mathscr{A}$ satisfies the conclusion of Theorem 2.5.2, then there exists a number N depending only on L, γ, and R (see inequality (2.113)) and such that, for any $i \in \mathbb{Z}$, one of the following alternatives is valid:*

$$\text{either } \Pi(i,N) > 2 \text{ or } \Pi(i+N,N) < 1/2. \tag{2.120}$$

Proof Fix numbers $i \in \mathbb{Z}$ and $N > 0$ and consider the sequence

$$w_k = -e_k, \quad i \le k \le i + 2N + 1.$$

It follows from the conclusion of Theorem 2.5.2 that there exists a sequence

$$\{v_k : i \le k \le i + 2N\}$$

such that

$$v_{k+1} = A_k v_k + w_{k+1} \text{ and } |v_k| \le L(2N+1)^\gamma, \quad i \le k \le i + 2N.$$

Set $v_k = a_k e_k$, where $a_k \in \mathbb{R}$. Then

$$a_{k+1} = \lambda_k a_k - 1 \text{ and } |a_k| \le L(2N+1)^\gamma, \quad i \le k \le i + 2N. \tag{2.121}$$

Now we show that there exists a large enough number N (depending only on L, γ, and R) such that if $a_{i+N} \ge 0$, then $\Pi(i,N) > 2$, and if $a_{i+N} < 0$, then $\Pi(i+N,N) < 1/2$.

Let us prove the existence of N for the first case (i.e., for the case where $a_{i+N} \ge 0$).

Since $\lambda_k > 0$, it follows from relations (2.121) that if $a_k \le 0$ for some $k \in [i, i+2N-1]$, then $a_{k+1} < 0$. Thus, if $a_{i+N} \ge 0$, then $a_i, \dots, a_{i+N-1} > 0$.

Relations (2.121) imply that in this case,

$$\lambda_k = \frac{a_{k+1}+1}{a_k}, \quad i \le k \le i + N - 1.$$

Hence,

$$\Pi(i,N) = \frac{a_{i+1}+1}{a_i}\frac{a_{i+2}+1}{a_{i+1}}\cdots\frac{a_{i+N}+1}{a_{i+N-1}} =$$

$$= \frac{1}{a_i}\frac{a_{i+1}+1}{a_{i+1}}\frac{a_{i+2}+1}{a_{i+2}}\cdots\frac{a_{i+N-1}+1}{a_{i+N-1}}(a_{i+N}+1) =$$

$$= \frac{a_{i+N}+1}{a_i}\prod_{k=i+1}^{i+N-1}\frac{a_k+1}{a_k},$$

and it follows from relations (2.121) that

$$\Pi(i,N) \geq \frac{1}{L(2N+1)^{\gamma}}\left(1+\frac{1}{L(2N+1)^{\gamma}}\right)^{N-1}. \tag{2.122}$$

Denote the expression on the right in (2.122) by $G_1(\gamma,N)$. Since

$$\log G_1(\gamma,N) = -\gamma\log(L(2N+1)) + (N-1)\log\left(1+\frac{1}{L(2N+1)^{\gamma}}\right),$$

$$\log\left(1+\frac{1}{L(2N+1)^{\gamma}}\right) \simeq (L(2N+1))^{-\gamma}$$

for large N, and $\gamma \in (0,1)$, we conclude that

$$G_1(\gamma,N) \to \infty, \quad N \to \infty.$$

Hence, there exists an N_1 depending only on L and γ such that $G_1(\gamma,N) > 2$ for $N \geq N_1$.

Now we consider the second case, i.e., we assume that $a_{i+N} < 0$. In this case, it follows from relations (2.121) that

$$a_k \in (-L(2N+1)^{\gamma}, -1), \quad i+N < k \leq i+2N. \tag{2.123}$$

As above, we set

$$\lambda_k = \frac{a_{k+1}+1}{a_k}.$$

Now we write

$$\Pi(i+N+1,N-1) = \frac{a_{i+N+2}+1}{a_{i+N+1}}\frac{a_{i+N+3}+1}{a_{i+N+2}}\cdots\frac{a_{i+2N}+1}{a_{i+2N-1}} =$$

$$= \frac{1}{a_{i+N+1}}\frac{a_{i+N+2}+1}{a_{i+N+2}}\frac{a_{i+N+3}+1}{a_{i+N+3}}\cdots\frac{a_{i+2N-1}+1}{a_{i+2N-1}}(a_{i+2N}+1)$$

and conclude that

$$\Pi(i+N+1,N-1) = \frac{a_{i+2N}+1}{a_{i+N+1}}\prod_{k=i+N+2}^{i+2N-1}\frac{a_k+1}{a_k}. \tag{2.124}$$

Inclusions (2.123) imply that

$$0 < \frac{a_k+1}{a_k} < 1 - \frac{1}{L(2N+1)^{\gamma}}, \quad i+N+2 \leq k \leq i+2N-1,$$

and

$$0 < \frac{a_{i+2N} + 1}{a_{i+N+1}} < L(2N + 1)^{\gamma}.$$

Combining these inequalities with (2.124), we conclude that

$$\Pi(i + N + 1, N - 1) < L(2N + 1)^{\gamma} \left(1 - \frac{1}{L(2N + 1)^{\gamma}}\right)^{N-2}.$$

Denote the right-hand side of the above inequality by $G_2(\gamma, N)$. Clearly, $G_2(\gamma, N) \to 0$ as $N \to \infty$; hence, there exists an N_2 depending only on L, γ, and R such that

$$G(\gamma, N) < \frac{1}{2R}, \quad N \geq N_2.$$

If $N \geq N_2$, then

$$\Pi(i + N, N) = \lambda_{i+N}\Pi(i + N + 1, N - 1) < R\frac{1}{2R} = 1/2.$$

Hence, the conclusion of our lemma holds for $N = \max(N_1, N_2)$.

□

Proof (of Theorem 2.5.3) Take an arbitrary $i \in \mathbb{Z}$ and the number N given by Lemma 2.5.1. The following statements hold:

(a) If $\Pi(i, N) > 2$, then $\Pi(i - N, N) > 2$;
(b) If $\Pi(i, N) < 1/2$, then $\Pi(i + N, N) < 1/2$.

Let us prove statement (a); the proof of statement (b) is similar.

By Lemma 2.5.1 applied to i–N, either $\Pi(i$–$N, N) > 2$ or $\Pi(i, N) < 1/2$. By the assumption of statement (a), the second case is impossible; thus, $\Pi(i - N, N) > 2$.

It follows from these statements that only one of the following cases is realized:

Case 1. $\Pi(i, N) > 2$ for all $i \in \mathbb{Z}$.
Case 2. $\Pi(i, N) < 1/2$ for all $i \in \mathbb{Z}$.
Case 3. There exist indices $i, j \in \mathbb{Z}$ such that $\Pi(i, N) > 2$ and $\Pi(j, N) < 1/2$.

Now we show that Theorem 2.5.3 is valid with $\mu = 2^{-1/N}$.

Consider Case 1. Take e_0 with $|e_0| = 1$ and define e_k, $k \in \mathbb{Z}$, by formulas (2.117). Represent any integer $l \geq 0$ in the form

$$l = nN + l_1, \quad n \in \mathbb{Z}_+, \; 0 \leq l_1 < N.$$

Then

$$\Pi(i,l) = \Pi(i,nN)\Pi(i+nN,l_1) > 2^n R^{-l_1}$$

(in the last estimate, we take into account inequalities (2.118)).

Hence, in Case 1,

$$\Pi(i,l) > R^{-l_1}\left(2^{-l_1/N}\right)\left(2^{1/N}\right)^l > C_0\mu^{-l}, \quad i \in \mathbb{Z},\ l \geq 0, \tag{2.125}$$

where

$$C_0 = R^{-N}/2.$$

Now we fix a point p_k of the trajectory $\{p_k\}$ and set $S(p_k) = \{0\}$ and $U(p_k)=\mathbb{R}$. Clearly, in this case, relations (2.114) and (2.115) are satisfied. Let us prove inequalities (2.116). Take any $v \in \mathbb{R} = U(p_k)$ and $l \geq 0$. Let

$$w = A_{k-l}^{-1} \cdots A_{k-1}^{-1} v.$$

Then

$$v = A_{k-1} \cdots A_{k-l} w.$$

Hence,

$$|v| = \lambda_{k-l} \cdots \lambda_{k-1} |w| = \Pi(k-l,l)|w|,$$

and it follows from (2.125) that

$$|w| \leq C\mu^l |v|,$$

where $C = (C_0)^{-1}$, as required.

In Case 2, we set $U(p_k) = \{0\}$ and $S(p_k) = \mathbb{R}$ and apply a similar reasoning.

Let us now consider Case 3. By our remark at the beginning of the proof,

$$\Pi(i-nN,N) > 2 \text{ and } \Pi(j+nN,N) < 1/2, \quad n \in \mathbb{Z}_+.$$

In this case, we set $S(p_k) = U(p_k) = \mathbb{R}$. Clearly, in this case, relation (2.114) is satisfied. Let us show how to prove inequalities (2.115).

We treat in detail two cases:

Case (I). $k + l \leq j$

and

Case (II). $k < j$ and $k + l > j$

(the remaining cases and the proof of inequalities (2.116) are left to the reader).

In Case (I), we note that $l \leq j - k$ and estimate

$$\Pi(k, l) \leq R^{j-k} = R^{j-k} 2^{l/N} 2^{-l/N} \leq C\mu^l,$$

where $C = R^{j-k} 2^{(j-k)/N}$. Hence,

$$|A_{k+l-1} \cdots A_k v| \leq C\mu^l |v|, \quad v \in S(p_k).$$

In Case (II), we represent $k + l = j + nN + l_1$, where $n \in \mathbb{Z}_+$ and $0 \leq l_1 < N$. Then

$$\Pi(k, l) = \Pi(k, j - k)\Pi(j, nN)\Pi(j + nN, l_1).$$

We note that $\Pi(k, j - k) \leq R^{j-k}$,

$$\Pi(j, nN) < 2^{-n} = 2^{l_1/N} \mu^l,$$

and

$$\Pi(j + nN, l_1) \leq R^{l_1} < R^N,$$

which gives us the desired estimate $\Pi(k, l) < C\mu^l$ (and, hence, inequalities (2.115)) with $C = 2R^{j-k+N}$. □

Historical Remarks Theorem 2.5.1 was published by S. B. Tikhomirov in [101].

Let us mention that earlier S. M. Hammel, J. A. Yorke, and C. Grebogi, based on results of numerical experiments, conjectured that a generic dissipative mapping $f : \mathbb{R}^2 \to \mathbb{R}^2$ belongs to a class $\mathrm{FHSP}_D(\mathcal{L}, C, d_0, 1/2, 1/2)$ [23, 24]. If this conjecture is true, then, in a sense, Theorem 2.5.1 cannot be improved.

2.6 A Homeomorphism with Lipschitz Shadowing and a Nonisolated Fixed Point

Consider the segment

$$I_0 = [-7/6, 4/3]$$

and a mapping $f_0 : I_0 \to I_0$ defined as follows:

$$f_0(x) = \begin{cases} 1 + (x - 1)/2, & x \in [1/3, 4/3]; \\ 2x, & x \in (-1/3, 1/3); \\ -1 + (x + 1)/2, & x \in [-7/6, -1/3]. \end{cases}$$

Clearly, the restriction f^* of f_0 to $[-1,1]$ is a homeomorphism of $[-1,1]$ having three fixed points: the points $x = \pm 1$ are attracting and the point $x = 0$ is repelling (and this homeomorphism f^* is an example of the so-called "North Pole – South Pole" dynamical system; every trajectory starting at a point $x \neq 0, \pm 1$ tends to an attractive fixed point as time tends to $+\infty$ and to the repelling fixed point as time tends to $-\infty$).

Now we define a homeomorphism $f : [-1,1] \to [-1,1]$. For an integer $n \geq 0$, denote $\mathcal{N}_n = 2^{-(n+2)}$ and set

$$f(x) = \mathcal{N}_n f_0(\mathcal{N}_n^{-1}(x - 3\mathcal{N}_n)) + 3\mathcal{N}_n, \quad x \in (2\mathcal{N}_n, 4\mathcal{N}_n]. \tag{2.126}$$

This defines f on $(0,1]$. Set $f(0) = 0$ and $f(x) = -f(-x)$ for $x \in [-1,0)$.

Clearly, f is a homeomorphism with a nonisolated fixed point $x = 0$ (for example, every point $x = \pm 2^{-n}$ is fixed). Let us note that in a neighborhood of any fixed point (with the exception of $x = 0$), f is either linearly expanding with coefficient 2 or linearly contracting with coefficient 1/2.

Theorem 2.6.1 *The homeomorphism f has the Lipschitz shadowing property.*

Before proving Theorem 2.6.1, we prove two auxiliary lemmas.

Lemma 2.6.1 *The mapping f_0 has the Lipschitz shadowing property on I_0.*

Proof Let

$$G_0 = (-1/3, 1/3)$$

and

$$G_1 = (-7/6, -1/3) \cup (1/3, 4/3).$$

We take d_0 small enough and $d \leq d_0$; in fact, we write below several explicit conditions on d and assume that they are satisfied.

There exist trivial cases where ξ is a subset of one of the segments $J_1 = [-7/6, -1/3]$, $J_2 = [-1/3, 1/3]$, or $J_3 = [1/3, 4/3]$.

Let, for example, $\xi \subset J_3$. The inequalities $1/3 \leq x_k \leq 4/3$ imply that

$$1/2 < 2/3 - d \leq f_0(x_k) - d \leq x_{k+1} \leq f_0(x_k) + d \leq 7/6 + d < 15/12.$$

These inequalities are satisfied for an arbitrary k; hence, ξ belongs to a domain in which f_0 is a hyperbolic diffeomorphism (and ξ is uniformly separated from the boundaries of the domain); by Theorem 1.4.2 (which, of course, is valid for infinite pseudotrajectories as well), there exist $\mathcal{L}, d_0 > 0$ such that if $d \leq d_0$, then ξ is $\mathcal{L}d$-shadowed by an exact trajectory of f_0.

A similar reasoning can be applied if $\xi \subset J_1$ or $\xi \subset J_2$.

To consider "nontrivial" cases, let us first describe possible positions of d-pseudotrajectories ξ of f_0 with small d with respect to $J_1, \ldots, J_3$.

First we show that such a pseudotrajectory cannot intersect both J_1 and J_3. Indeed, if we assume that $\xi \cap J_3 \neq \emptyset$, i.e., there exists an index m such that $x_m \geq 1/3$, then

$$x_{m+1} \geq f_0(1/3) - d = 2/3 - d > 1/3$$

and, consequently,

$$x_{m+i} > 1/3, \quad i > 0.$$

Similarly, if there exists an index l such that $x_l \leq -1/3$, then

$$x_{l+1} \leq -2/3 + d < -1/3$$

and

$$x_{l+i} < -1/3, \quad i > 0,$$

and we get a contradiction.

Thus, it remains to consider the cases where either

$$\xi \subset J_2 \cup J_3, \quad \xi \cap \mathrm{Int}(J_2) \neq \emptyset, \quad \xi \cap \mathrm{Int}(J_3) \neq \emptyset,$$

or

$$\xi \subset J_1 \cup J_2, \quad \xi \cap \mathrm{Int}(J_1) \neq \emptyset, \quad \xi \cap \mathrm{Int}(J_2) \neq \emptyset.$$

We consider the first case; the reasoning in the second case is similar.

We claim that in the case considered, ξ contains two points x_k, x_l such that

$$0 < x_k < 1/3 < x_l. \tag{2.127}$$

The existence of the point x_l follows directly from our assumption; it is easily seen that

$$x_{l+i} \geq 2/3 - d > 1/2, \quad i > 0. \tag{2.128}$$

Thus, either the set

$$\{m : x_m \in \mathrm{Int}(J_2), x_m \leq 0\}$$

is empty (which implies that there exists an index k for which inequality (2.127) is valid) or it is nonempty and bounded from above. In the latter case, let m_0 be its maximal element. Then

$$x_{m_0+1} \leq f_0(x_{m_0}) + d \leq d$$

(i.e., $x_{m_0+1} \in J_2$) and $x_{m_0+1} > 0$; thus, we get the required $k = m_0 + 1$.

Obviously, $l > k$ (see (2.128)). Consider the finite set of indices

$$\kappa = \{ i \in [k, l-1] : x_i \le 1/3 \}.$$

This set is nonempty ($k \in \kappa$) and finite; hence, it contains the maximal element. Let it be x_{k_0}; clearly,

$$x_{k_0} \le 1/3 < x_{k_0+1}.$$

To simplify notation, let us assume that $k_0 = 0$. Thus,

$$x_0 \le 1/3 < x_1.$$

In this case,

$$x_i \ge 2/3 - d > 1/2, \quad i \ge 2. \tag{2.129}$$

On the other hand,

$$x_1 \le 2/3 + d < 1,$$

and one easily shows that

$$x_i \le 1 + 2d, \quad i \ge 2. \tag{2.130}$$

Since f_0^{-1} has Lipschitz constant 2, ξ is a $2d$-pseudotrajectory of f_0^{-1}; hence,

$$x_{-1} \le 1/6 + 2d < 2/9,$$

and, applying the same reasoning as above, we conclude that

$$-4d < x_i < 1/6 + 2d < 2/9, \quad i < 0. \tag{2.131}$$

Now we show that there exists a d_0 such that if $d \le d_0$ and $p = x_0$, then

$$\left| f_0^k(p) - x_k \right| < 3d, \quad k \in \mathbb{Z}. \tag{2.132}$$

First, clearly,

$$|f_0(p) - x_1| < d.$$

Since the Lipschitz constant of f_0 is 2,

$$\left| f_0^2(p) - x_2 \right| \le |f(f(p)) - f(x_1)| + |f(x_1) - x_2| < 2d + d = 3d.$$

It follows from (2.129) that

$$f_0^2(p) > 1/2 - 3d > 1/3,$$

and then

$$f_0^k(p) > 1/3, \quad k \geq 2.$$

Hence,

$$\left|f_0^3(p) - x_3\right| \leq \left|f_0(f_0^2(p)) - f_0(x_2)\right| + |f_0(x_2) - x_3| < 3d/2 + d < 3d.$$

Repeating these estimates, we establish inequalities (2.132) for $k \geq 2$.

On the other hand, the inclusion $p \in J_2$ implies that $f_0^k(p) \in J_2$ for $k \leq 0$. Since $f_0^{-1}(x) = x/2$ for $x \in J_2$ and (2.131) holds, the inequality

$$|f_0(x_1) - p| < d$$

implies that

$$|x_1 - f_0^{-1}(p)| < d/2.$$

After that, we estimate

$$\left|x_2 - f_0^{-2}(p)\right| \leq \left|x_2 - f_0^{-1}(x_{-1})\right| + \left|f_0^{-1}(x_{-1}) - f_0^{-1}(f_0^{-1}(p))\right| < d/2 + d/2,$$

and so on, which shows that an analog of (2.132) with $3d$ replaced by d holds for $k < 0$. □

The following statement is almost obvious.

Lemma 2.6.2 *Let g be a mapping of a segment J and let numbers $M > 0$ and m be given. Consider the mapping*

$$g'(y) = M^{-1}g(M(y - m)) + m$$

on the set

$$J' = \{y : \; M(y - m) \in J\}.$$

If g has the Lipschitz shadowing property with constants $\mathcal{L}, d_0$, then g' has the Lipschitz shadowing property with constants $\mathcal{L}, M^{-1}d_0$.

Proof First we note that if $\{y_k\}$ is a d-pseudotrajectory of g' with $d \leq d_0/M$ and $x_k = M(y_k - m)$, then

$$g(x_k) - x_{k+1} = M(g'(y_k) - y_{k+1}).$$

Hence, $\{x_k\}$ is an Md-pseudotrajectory of g.

Since $Md \leq d_0$, there exists a point p such that

$$|g^k(p) - x_k| \leq \mathscr{L}Md.$$

Set $p' = M^{-1}p + m$. Then, obviously,

$$|(g')^k(p') - y_k| = M^{-1}|g^k(p) - x_k| \leq \mathscr{L}d.$$

□

Let us prove Theorem 2.6.1.

Proof For a natural n, define the segment

$$I_n = [\alpha_n, \beta_n] = [11\mathscr{N}_n/6, 13\mathscr{N}_n/3]$$

and note that formula (2.126) defining f for $x \in (2\mathscr{N}_n, 4\mathscr{N}_n]$ is, in fact, valid for $x \in I_n$.

To prove Theorem 2.6.1, we first claim that there exists a constant c independent of n such that if d satisfies a condition of the form

$$d \leq c\mathscr{N}_n \tag{2.133}$$

and $\xi = \{x_k\}$ is a d-pseudotrajectory of f that intersects I_n, then ξ is a subset of one of the segments I_{n-1}, I_n, I_{n+1}.

In fact, all the conditions imposed below on d have the form (2.133).

It follows from the inequalities

$$f(\alpha_n) = 23\mathscr{N}_n/12 > \alpha_n, \quad f(\beta_n) = 25\mathscr{N}_n/6 < \beta_n$$

that if c is small enough (we do not repeat this assumption below), then

$$\mathrm{Cl}(N(d, f(I_m))) \subset I_m, \quad m = n-1, n, n+1. \tag{2.134}$$

Thus, if $x_k \in I_m$ for some $m = n-1, n, n+1$, then it follows from (2.134) that

$$x_{k+i} \in I_m, \quad i \geq 0. \tag{2.135}$$

Let $x_0 \in I_n$.

We assume that

$$\mathrm{Cl}(N(2d, f^{-1}(I_n))) \subset I_{n-1} \cup I_n \cup I_{n+1}$$

(note that this condition on d has precisely form (2.133)).

By (2.135), $x_k \in I_n$ for $k \geq 0$. Thus, if the inclusion $\xi \subset I_n$ does not hold, there exists an index $l < 0$ such that

$$x_l \in (I_{n-1} \cup I_{n+1}) \setminus I_n$$

(recall that ξ is a $2d$-pseudotrajectory of f^{-1}).

Assume, for definiteness, that $x_l \in I_{n-1}$ (the remaining case is treated similarly). In this case, the same inclusions (2.135) imply that

$$x_{l+i} \in I_{n-1}, \quad i \geq 0.$$

To show that

$$x_{l+i} \in I_{n-1}, \quad i < 0,$$

take an index $m < l$ and assume that $x_m \in I_\nu$. Then inclusions (2.135) imply that

$$x_0, x_l \in I_\nu;$$

hence,

$$I_\nu \cap I_n \neq \emptyset \text{ and } I_\nu \cap I_{n-1} \neq \emptyset,$$

from which it follows that either $\nu = n$ or $\nu = n-1$. But since $x_l \notin I_n$, $\nu \neq n$, and we conclude that $\xi \subset I_{n-1}$, as claimed.

Of course, a similar statement holds for the segments $I'_n = [-\beta_n, -\alpha_n]$.

Without loss of generality, we assume that

$$c \leq d_0/2, \tag{2.136}$$

where d_0 is given by Lemma 2.6.1. Let $\delta(m) = c\mathcal{N}_m$.

Consider a d-pseudotrajectory $\xi = \{x_k\} \subset [-1, 1]$ of f with $d \leq d_0$. If

$$d \geq \delta(0) = c\mathcal{N}_0 = c/4,$$

then $1 \leq 4d/c$, and ξ is $4d/c$-shadowed by the fixed point $x = 0$.

Otherwise, we find the maximal index m_0 for which $d < \delta(m_0)$. In this case,

$$d \geq \delta(m_0 + 1) = \delta(m_0)/2. \tag{2.137}$$

First we assume that

$$\xi \cap I_m \neq \emptyset \text{ for some } m \leq m_0 \tag{2.138}$$

(the case of I'_m is similar).

In this case, the inequalities

$$d < \delta(m_0) \leq \delta(m)$$

imply that ξ is a subset of one of the segments I_{m-1}, I_m, I_{m+1}. We assume that $\xi \subset I_{m+1}$; in the remaining cases, the same estimates work.

Since

$$d \leq \delta(m) = c\mathcal{N}_m \leq d_0\mathcal{N}_m/2 = d_0\mathcal{N}_{m+1}$$

(we refer to (2.136)), Lemma 2.6.2 implies that ξ is $\mathcal{L}$-shadowed.

If relation (2.138) does not hold, then

$$|x_k| \leq \alpha_{m_0} = \frac{11N_{m_0}}{6} = \frac{11\delta(m_0)}{6c} \leq \frac{11}{3c}d$$

(we take into account inequality (2.137) in the last estimate). Thus, in this case, ξ is $11d/(3c)$-shadowed by the fixed point $x = 0$. □

Historical Remarks In this section, we give a simplified proof of Theorem 2.6.1 compared to the original variant published by A. A. Petrov and the first author in [59].

2.7 Lipschitz Shadowing Implies Structural Stability: The Case of a Vector Field

Let M be a smooth closed manifold with Riemannian metric dist and let X be a vector field on M of class C^1. Denote by $\phi(t, x)$ the flow on M generated by the vector field X.

Our main goal in this section is to prove the following statement.

Theorem 2.7.1 *If a vector field X has the Lipschitz shadowing property, then X is structurally stable.*

In the proof of Theorem 2.7.1, we refer to Theorem 1.3.14.

Define a diffeomorphism f on M by setting $f(x) = \phi(1, x)$.

It is an easy exercise to show that the chain recurrent set $\mathcal{R}(\phi)$ of the flow ϕ (see Definition 1.3.22) coincides with the chain recurrent set of the diffeomorphism f.

2.7.1 Discrete Lipschitz Shadowing for Flows

In this section, we introduce the notion of discrete Lipschitz shadowing for a vector field in terms of the diffeomorphism $f(x) = \phi(1, x)$ introduced above and show that the Lipschitz shadowing property of ϕ implies the discrete Lipschitz shadowing.

Definition 2.7.1 A vector field X has the *discrete Lipschitz shadowing property* if there exist $d_0, L > 0$ such that if $y_k \in M$ is a sequence with

$$\mathrm{dist}(y_{k+1}, f(y_k)) \leq d \leq d_0, \quad k \in \mathbb{Z}, \tag{2.139}$$

then there exist sequences $x_k \in M$ and $t_k \in \mathbb{R}$ such that

$$|t_k - 1| \leq Ld, \ \mathrm{dist}(x_k, y_k) \leq Ld, \ x_{k+1} = \phi(t_k, x_k), \quad k \in \mathbb{Z}. \tag{2.140}$$

Lemma 2.7.1 *The Lipschitz shadowing property of ϕ implies the discrete Lipschitz shadowing of X.*

Proof First we note that since M is compact and X is C^1-smooth, there exists a $\nu > 0$ such that

$$\mathrm{dist}(\phi(t, x), \phi(t, y)) \leq \nu \mathrm{dist}(x, y), \quad x, y \in M, \ t \in [0, 1]. \tag{2.141}$$

Consider a sequence y_k that satisfies inequalities (2.139) and define a mapping $y : \mathbb{R} \to M$ by setting

$$y(t) = \phi(t - k, y_k), \quad k \leq t < k + 1, \ k \in \mathbb{Z}.$$

Fix a $\tau \in [k, k + 1)$. If $t \in [0, 1]$ and $\tau + t < k + 1$, then

$$\mathrm{dist}\,(y(\tau + t), \phi(t, y(\tau))) = \mathrm{dist}\,(\phi(\tau + t - k, y_k), \phi(t, \phi(\tau - k, y_k))) = 0.$$

If $k + 1 \leq \tau + t$, then

$$\mathrm{dist}(y(\tau + t), \phi(t, y(\tau))) = \mathrm{dist}(\phi(\tau + t - k - 1, y_{k+1}), \phi(\tau + t - k, y_k)) =$$

$$= \mathrm{dist}(\phi(\tau + t - k - 1, y_{k+1}), \phi(\tau + t - k - 1, \phi(1, y_k))) \leq \nu d.$$

Thus, $y(t)$ is a $(\nu + 1)d$-pseudotrajectory of ϕ. Hence, if $d \leq d_0/(\nu + 1)$, where d_0 is from the definition of the Lipschitz shadowing property for ϕ, then there exists a trajectory $x(t)$ of X and a reparametrization

$$\alpha \in \mathrm{Rep}(\mathscr{L}(\nu + 1)d)$$

such that

$$\operatorname{dist}(y(t), x(\alpha(t))) \leq \mathscr{L}(\nu + 1)d, \quad t \in \mathbb{R}.$$

If we set

$$x_k = x(\alpha(k)) \text{ and } t_k = \alpha(k+1) - \alpha(k),$$

then

$$x_{k+1} = x(\alpha(k+1)) = \phi(\alpha(k+1) - \alpha(k), x(\alpha(k)) = \phi(t_k, x_k),$$

$$\operatorname{dist}(x_k, y_k) = \operatorname{dist}(x(\alpha(k)), y_k) \leq \mathscr{L}(\nu + 1)d,$$

and

$$|t_k - 1| = \left| \frac{\alpha(k+1) - \alpha(k)}{k + 1 - k} - 1 \right| \leq \mathscr{L}(\nu + 1)d.$$

Taking $L = \mathscr{L}(\nu + 1)$ and d_0 in Definition 2.7.1 as $d_0/(\nu + 1)$, we complete the proof of the lemma. □

As in Sect. 2.3, we reduce our shadowing problem to the problem of existence of bounded solutions of certain difference equations. To clarify the presentation, we again first take $M = \mathbb{R}^n$, assume that the considered vector field X defines a flow (every trajectory is defined for $t \in \mathbb{R}$), and assume that the diffeomorphism f satisfies Condition S formulated in Sect. 2.3 (see estimate (2.52)). To treat the general case of a compact manifold M, one has to apply exponential mappings (see Remark 2.7.1 below); we leave details to the reader.

As above, we denote

$$\|V\| = \sup_{k \in \mathbb{Z}} |v_k|$$

for a bounded sequence of vectors $V = \{v_k : k \in \mathbb{Z}\}$.

Lemma 2.7.2 *Assume that X has the discrete Lipschitz shadowing property with constant L. Let $x(t)$ be an arbitrary trajectory of X, let $p_k = x(k)$, and set $A_k = Df(p_k)$ (recall that $f(x) = \phi(1, x)$). Assume that f satisfies Condition S formulated in Sect. 2.3. Let $B = \{b_k \in \mathbb{R}^n\}$ be a bounded sequence and denote $\beta_0 = \|B\|$.*

Then there exists a sequence of scalars s_k with

$$|s_k| \leq \beta = L(\beta_0 + 1)$$

such that the difference equation

$$v_{k+1} = A_k v_k + X(p_{k+1})s_k + b_{k+1} \tag{2.142}$$

has a solution $V = \{v_k\}$ *with*

$$\|V\| \le \beta. \tag{2.143}$$

Proof Fix a natural number N and define $\Delta_k \in \mathbb{R}^n$ as the solution of

$$v_{k+1} = A_k v_k + b_{k+1}, \quad k = -N, \dots, N-1,$$

with $\Delta_{-N} = 0$. Then

$$|\Delta_k| \le C, \quad k = -N, \dots, N, \tag{2.144}$$

where C depends on N, β_0, and an upper bound of $\|A_k\|$ for $k = -N, \dots, N-1$.

Fix a small number $d > 0$ and fix μ in (2.52) so that

$$\mu C < 1. \tag{2.145}$$

Consider the sequence of points $y_k \in \mathbb{R}^n$ defined as follows: $y_k = p_k$ for $k \le -N$,

$$y_k = p_k + d\Delta_k, \quad k = -N, \dots, N-1,$$

and $y_{N+k} = f^k(y_N)$ for $k > 0$.

Then $y_{k+1} = f(y_k)$ for $k \le -N-1$ and $k \ge N$.

Since

$$y_{k+1} = p_{k+1} + d\Delta_{k+1} = p_{k+1} + dA_k\Delta_k + db_{k+1},$$

$$|y_{k+1} - p_{k+1} - dA_k\Delta_k| \le d|b_{k+1}| \le d\beta_0. \tag{2.146}$$

On the other hand, if $dC \le \delta(\mu)$, then it follows from (2.52) that

$$|f(y_k) - p_{k+1} - dA_k\Delta_k| = |f(p_k + d\Delta_k) - f(p_k) - dA_k\Delta_k| \le$$

$$\le \mu|d\Delta_k| \le \mu dC < d \tag{2.147}$$

(see (2.145)).

Combining (2.146) and (2.147), we see that

$$|y_{k+1} - f(y_k)| < d(\beta_0 + 1), \quad k \in \mathbb{Z},$$

if d is small enough. Let us emphasize that the required smallness of d depends on β_0, N, and estimates on $\|A_k\|$.

Now the assumptions of our lemma imply that there exist sequences x_k and t_k such that

$$|t_k - 1| \le d\beta, \ |x_k - y_k| \le d\beta, \ x_{k+1} = \phi(t_k, x_k), \quad k \in \mathbb{Z}.$$

If we represent

$$x_k = p_k + dc_k \text{ and } t_k = 1 + ds_k,$$

then

$$|dc_k - d\Delta_k| = |x_k - y_k| \le d\beta.$$

Thus,

$$|c_k - \Delta_k| \le \beta, \quad -N \le k \le N. \tag{2.148}$$

Clearly,

$$|s_k| \le \beta, \quad k \in \mathbb{Z}. \tag{2.149}$$

Define mappings

$$G_k : \mathbb{R} \times \mathbb{R}^n \to \mathbb{R}^n, \quad k \in \mathbb{Z},$$

by

$$G_k(t, v) = \phi(1 + t, p_k + v) - p_{k+1}.$$

Then

$$G_k(0,0) = 0, \ D_t G_k(t, v)|_{t=0,v=0} = X(p_{k+1}), \ D_v G_k(t, v)|_{t=0,v=0} = A_k.$$

We can write the equality

$$x_{k+1} = \phi(1 + ds_k, x_k)$$

in the form

$$p_{k+1} + dc_{k+1} = \phi(1 + ds_k, p_k + dc_k),$$

which is equivalent to

$$dc_{k+1} = G_k(ds_k, dc_k). \tag{2.150}$$

Now we fix a sequence of values $d = d^{(m)} \to 0,\ m \to \infty$. Let us denote by $c_k^{(m)}$, $t_k^{(m)}$, and $s_k^{(m)}$ the values c_k, t_k, and s_k defined above and corresponding to $d = d^{(m)}$.

It follows from estimates (2.148) and (2.149) that $|c_k^{(m)}| \leq C + \beta$ and $|s_k^{(m)}| \leq \beta$ for all m and $-N \leq k \leq N-1$. The second inequality implies that $\left|t_k^{(m)}\right| \leq 1$ for large m. Hence (passing to a subsequence, if necessary), we can assume that

$$c_k^{(m)} \to \tilde{c}_k,\ t_k^{(m)} \to \tilde{t}_k,\ s_k^{(m)} \to \tilde{s}_k, \quad m \to \infty,$$

for $-N \leq k \leq N-1$.

Applying relations (2.150) and (2.149), we can write

$$d_m c_{k+1}^{(m)} = G_k\left(d_m s_k^{(m)}, d_m c_k^{(m)}\right) = A_k d_m c_{k+1}^{(m)} + X(p_{k+1}) d_m s_k^{(m)} + o(d_m).$$

Dividing these equalities by d_m, we get the relations

$$c_{k+1}^{(m)} = A_k c_{k+1}^{(m)} + X(p_{k+1}) s_k^{(m)} + o(1), \quad -N \leq k \leq N-1.$$

Letting $m \to \infty$, we arrive at the relations

$$\tilde{c}_{k+1} = A_k \tilde{c}_k + X(p_{k+1}) \tilde{s}_k, \quad -N \leq k \leq N-1,$$

where

$$|\Delta_k - \tilde{c}_k|, |\tilde{s}_k| \leq \beta, \quad -N \leq k \leq N-1,$$

due to (2.148) and (2.149).

Recall that N was fixed in the above reasoning. Denote the obtained $\tilde{s}_k$ by $s_k^{(N)}$. Then $v_k^{(N)} = \Delta_k - \tilde{c}_k$ is a solution of the equations

$$v_{k+1}^{(N)} = A_k v_k^{(N)} + X(p_{k+1}) s_k^{(N)} + b_{k+1}, \quad -N \leq k \leq N-1,$$

such that $\left|v_k^{(N)}\right| \leq \beta$.

There exist subsequences $v_k^{(j_N)} \to v_k'$ and $s_k^{(j_N)} \to s_k'$ as $N \to \infty$ (we do not assume uniform convergence) such that

$$v_{k+1}' = A_k v_k' + X(p_{k+1}) s_k' + b_{k+1}, \quad k \in \mathbb{Z},$$

and $|v_k'|, |s_k'| \leq \beta$. The lemma is proved. □

Remark 2.7.1 An analog of Lemma 2.7.2 is valid in the case of a smooth closed manifold M. In this case, we denote $\mathcal{M}_k = T_{p_k}M$ and consider the difference equation (2.142) in which $v_k, b_k \in \mathcal{M}_k$.

Proving an analog of Lemma 2.7.2 in the case of a closed manifold (and replacing, for example, the formula $y_k = p_k + d\Delta_k$ by $y_k = \exp_{p_k}(d\Delta_k)$, compare with the proof of Lemma 2.3.3 in Sec 2.3), one gets a similar statement with the estimates $|s_k| \leq \beta := L(2\beta_0 + 1)$ and $\|V\|_\infty \leq 2\beta$ (see the original paper [57]).

Thus, in what follows, we refer to Lemma 2.7.2 in the case of a vector field X on a smooth closed manifold M (with $B = \{b_k \in \mathbb{R}^n\}$ replaced by $B = \{b_k \in \mathscr{M}_k\}$ and properly corrected estimates).

2.7.2 Rest Points

In this section, we show that if a vector field has the Lipschitz shadowing property, then its rest points are hyperbolic and isolated in the chain recurrent set. Thus, in what follows we assume that we work with a vector field X on a smooth closed manifold M having the Lipschitz shadowing property.

Lemma 2.7.3 *Every rest point of X is hyperbolic.*

Proof Let x_0 be a rest point. Applying an analog of Lemma 2.7.2 for the case of a manifold with $p_k = x_0$ and noting that $X(p_k) = 0$, we conclude that the difference equation

$$v_{k+1} = Df(x_0)v_k + b_{k+1}$$

has a bounded solution for any bounded sequence $b_k \in \mathscr{M}_k$ (recall that $\mathscr{M}_k = T_{p_k}M$).

Then it follows from the Maizel' theorem (see Theorem 2.1.1 of Sect. 2.1) that the constant sequence $\mathscr{A} = \{A_k = Df(x_0)\}$ is hyperbolic on $\mathbb{Z}_+$; in particular, every bounded solution of the equation

$$v_{k+1} = Df(x_0)v_k$$

tends to 0 as $k \to \infty$.

However, if the rest point x_0 is not hyperbolic, then the matrix $Df(x_0)$ has an eigenvalue on the unit circle, in which case the above equation has a nontrivial solution with constant norm. Thus, x_0 is hyperbolic. □

Lemma 2.7.4 *Rest points are isolated in the chain recurrent set $\mathscr{R}(\phi)$.*

Proof Let us assume that there exists a rest point x_0 that is not isolated in $\mathscr{R}(\phi)$. First we want to show that there is a homoclinic trajectory $x(t)$ associated with x_0.

Since x_0 is hyperbolic by the previous lemma, there exists a small $d > 0$ and a number $a > 0$ such that if $\operatorname{dist}(\phi(t, y), x_0) \leq \mathscr{L}d$ for $|t| \geq a$, then $\phi(t, y) \to x_0$ as $|t| \to \infty$.

Assume that there exists a point $y \in \mathscr{R}(\phi)$ such that y is arbitrarily close to x_0 and $y \neq x_0$. Given any $\varepsilon_0, \theta > 0$ we can find points $y_1, \ldots, y_N$ and numbers

$T_0, \dots, T_N > \theta$ such that

$$\operatorname{dist}(\phi(T_0, y), y_1) < \varepsilon_0,$$

$$\operatorname{dist}(\phi(T_i, y_i), y_{i+1}) < \varepsilon_0, \quad i = 1, \dots, N,$$

and

$$\operatorname{dist}(\phi(T_N, y_N), y_1) < \varepsilon_0.$$

Set $T = T_0 + \cdots + T_N$ and define g^* on $[0, T]$ by

$$g^*(t) = \begin{cases} \phi(t, y), \; 0 \le t \le T_0; \\ \phi(t, y_i), \; T_0 + \cdots + T_{i-1} < t < T_0 + \cdots + T_i; \\ y, \quad t = T. \end{cases}$$

Clearly, for any $\varepsilon > 0$ we can find ε_0 depending only on ε and ν (see (2.141)) such that $g^*(t)$ is an ε-pseudotrajectory of ϕ on $[0, T]$.

Now we define

$$g(t) = \begin{cases} x_0, \quad t \le 0; \\ g^*(t), \; 0 < t \le T; \\ x_0, \quad t > T. \end{cases}$$

We want to choose y and ε in such a way that $g(t)$ is a d-pseudotrajectory of ϕ. We have to show that

$$\operatorname{dist}(\phi(t, g(\tau)), g(t + \tau)) < d \tag{2.151}$$

for all τ and $t \in [0, 1]$.

Clearly, (2.151) holds for (i) $\tau \le -1$, (ii) $\tau \ge T$, (iii) $\tau, \tau + t \in [-1, 0]$, and (iv) $\tau, \tau + t \in [0, T]$ and $\varepsilon < d$.

If $\tau \in [-1, 0]$ and $\tau + t > 0$, then

$$\operatorname{dist}(\phi(t, g(\tau)), g(t + \tau)) = \operatorname{dist}(x_0, g^*(t + \tau)) \le$$

$$\le \operatorname{dist}(x_0, \phi(t + \tau, y)) + \operatorname{dist}(\phi(t + \tau, y), g^*(t + \tau)) \le \nu \operatorname{dist}(x_0, y) + \varepsilon,$$

where ν is as in (2.141). The last value is less than d if $\operatorname{dist}(x_0, y)$ and ε are small enough. Note that, for a fixed y, we can decrease ε and increase $N, T_0, \dots, T_N$ arbitrarily so that $g(t)$ remains a d-pseudotrajectory.

Similarly, (2.151) holds if $\tau \in [0, T]$ and $\tau + t > T$.

Thus, $g(t)$ is $\mathcal{L}d$ shadowed by a trajectory $x(t)$ such that $\operatorname{dist}(x(t), x_0) \le \mathcal{L}d$ if $|t|$ is sufficiently large; hence, $x(t) \to x_0$ as $|t| \to \infty$.

Now we want to show that $x(t)$ is a homoclinic trajectory if d is small enough. For this purpose, we have to show that $x(t) \neq x_0$.

There exists an $\varepsilon_1 > \mathscr{L}d$ (provided that d is small enough) such that if y does not belong to the local stable manifold of x_0, then $\text{dist}(\phi(t_0), y) \geq \varepsilon_1$ for some $t_0 > 0$. We can choose $T_0 > t_0$ (not changing the point y). Then $g(t)$ contains the point $g^*(t_0) = \phi(t_0, y)$ whose distance to x_0 is more than $\mathscr{L}d$. Hence, $x(t)$ contains a point different from x_0, as was claimed.

If y belongs to the local stable manifold of x_0, then it does not belong to the local unstable manifold of x_0. In this case, considering the flow $\psi(t, x) = \phi(-t, x)$, we can apply the above reasoning to ψ noting that $\mathscr{R}(\psi) = \mathscr{R}(\phi)$ and ψ has the Lipschitz shadowing property as well.

Now we show that the existence of this homoclinic trajectory $x(t)$ leads to a contradiction. Set $p_k = x(k)$. Since $A_k X(p_k) = X(p_{k+1})$, it is easily verified that if we consider two sequences β_k and s_k such that

$$\beta_{k+1} = \beta_k + s_k, \quad k \in \mathbb{Z},$$

then $u_k = \beta_k X(p_k)$ is a solution of

$$u_{k+1} = A_k u_k + X(p_{k+1}) s_k, \quad k \in \mathbb{Z}. \tag{2.152}$$

In addition, if the sequence s_k is bounded, then the sequence $\beta_k X(p_k)$ is bounded as well since $X(p_k) \to 0$ exponentially as $|k| \to \infty$ (the trajectory $x(t)$ tends to a hyperbolic rest point as time goes to $\pm\infty$) and the sequence $|\beta_k|/|k|$ is bounded).

By Lemma 2.7.2, for any bounded sequence $b_k \in \mathscr{M}_k$ there exists a bounded scalar sequence s_k such that Eqs. (2.142) have a bounded solution v_k. We have shown that Eqs. (2.152) have a bounded solution u_k. Then the sequence $w_k = v_k - u_k$ is bounded and satisfies the equations

$$w_k = A_k w_k + b_{k+1}, \quad k \in \mathbb{Z}.$$

Thus, the sequence $\mathscr{A} = \{A_k\}$ has the Perron property on $\mathbb{Z}$. It follows from Theorems 2.1.1 and 2.1.2 that the sequence $\mathscr{A}$ is hyperbolic both on $\mathbb{Z}_+$ and $\mathbb{Z}_-$ and the corresponding spaces S_0^+ and U_0^- are transverse. But this leads to a contradiction since

$$\dim S_0^+ + \dim U_0^- = \dim M$$

(because $\dim S_0^+$ equals the dimension of the stable manifold of the hyperbolic rest point x_0 and $\dim U_0^-$ equals the dimension of its unstable manifold), while any of the spaces S_0^+ and U_0^- contains the nonzero vector $X(p_0)$. The lemma is proved. □

2.7.3 Hyperbolicity of the Chain Recurrent Set

We have shown that rest points of ϕ are hyperbolic and isolated in the chain recurrent set $\mathscr{R}(\phi)$. Since M is compact, this implies that the set $\mathscr{R}(\phi)$ is the union of a finite set of hyperbolic rest points and a compact set (let us denote it Σ) on which the vector field X is nonzero.

To show that $\mathscr{R}(\phi)$ is hyperbolic, it remains to show that the set Σ is hyperbolic.

Consider the subbundle $\mathscr{V}(\Sigma)$ of the tangent bundle $TM|_\Sigma$ defined in Sect. 1.3 before Theorem 1.3.15.

Let $x(t)$ be a trajectory in Σ. Let us introduce the following notation. Put $p_k = x(k)$ and let $P_k = P(p_k)$ and $V_k = V(p_k)$ (recall that $P(x)$ is the orthogonal projection in T_xM with kernel spanned by $X(x)$ and $V(x)$ is the orthogonal complement to $X(x)$ in T_xM). Introduce the operators

$$B_k = P_{k+1}A_k : \ V_k \to V_{k+1}$$

(recall that $A_k = Df(p_k)$).

Lemma 2.7.5 *For every bounded sequence $b_k \in V_k$ there exists a bounded solution $v_k \in V_k$ of*

$$v_{k+1} = B_k v_k + b_{k+1}, \quad k \in \mathbb{Z}. \tag{2.153}$$

Proof Fix a bounded sequence $b_k \in V_k$. There exist bounded sequences s_k of scalars and w_k of vectors in $T_{p_k}M$ such that

$$w_{k+1} = A_k w_k + X(p_{k+1})s_k + b_{k+1}, \quad k \in \mathbb{Z}, \tag{2.154}$$

(see the remark after Lemma 2.7.2).

Note that $A_k X(p_k) = X(p_{k+1})$. Since $(\mathrm{Id} - P_k)v \in \{X(p_k)\}$ for $v \in \mathscr{M}_k$, we see that

$$P_{k+1}A_k(\mathrm{Id} - P_k) = 0,$$

which gives us the equality

$$P_{k+1}A_k = P_{k+1}A_k P_k. \tag{2.155}$$

The properties of the set Σ imply that the projections P_k are uniformly bounded.

Multiplying (2.154) by P_{k+1}, taking into account the equalities $P_{k+1}X(p_{k+1}) = 0$ and $P_{k+1}b_{k+1} = b_{k+1}$, and applying (2.155), we conclude that $v_k = P_k w_k$ is the required bounded solution of (2.153). The lemma is proved. □

It follows from the above lemma that if we fix a trajectory $x(t)$ in Σ and consider the corresponding sequence of operators $\mathscr{B} = \{B_k\}$, then $\mathscr{B}$ has the Perron property.

By Theorems 2.1.1 and 2.1.2, the sequence $\mathscr{B}$ is hyperbolic both on $\mathbb{Z}_-$ and $\mathbb{Z}_+$ and the corresponding spaces $U_0^-(\mathscr{B})$ and $S_0^+(\mathscr{B})$ are transverse.

Consider the mapping π on the normal bundle $\mathscr{V}(\Sigma)$ defined in Sect. 1.3. Recall that

$$\pi(x, v) = (f(x), B(x)v), \text{ where } B(x) = P(f(x))Df(x)$$

(see Sect. 1.3).

In fact, we have shown that π satisfies an analog of the strong transversality condition.

The same reasoning as in the proof of Lemma 2.2.5 shows that the dual mapping π^* does not have nontrivial bounded trajectories. It is easy to show that if the flow ϕ has the shadowing property, then its nonwandering set coincides with its chain recurrent set.

Hence, we can repeat the reasoning of the proof of Theorem 2.2.2 to conclude that the mapping π is hyperbolic.

It remains to refer to Theorem 1.3.15 to conclude that Σ is a hyperbolic set of the flow ϕ.

2.7.4 *Transversality of Stable and Unstable Manifolds*

Let $x(t)$ be a trajectory that belongs to the intersection of the stable and unstable manifolds of two trajectories, $x_+(t)$ and $x_-(t)$, respectively, lying in the chain recurrent set of ϕ.

Without loss of generality, we may assume that

$$\text{dist}(x(0), x_+(0)) \to 0, \quad t \to \infty,$$

and

$$\text{dist}(x(0), x_-(0)) \to 0, \quad t \to -\infty.$$

Denote $p_k = x(k)$, $k \in \mathbb{Z}$; let $W^s(p_0)$ and $W^u(p_0)$ be the stable and unstable manifolds of p_0, respectively. Then, of course, $W^s(p_0) = W^s(x_+(0))$ and $W^u(p_0) = W^u(x_-(0))$. Denote by E^s and E^u the tangent spaces of $W^s(p_0)$ and $W^u(p_0)$ at p_0.

We use the notation introduced before Lemma 2.7.5.

By Lemma 2.7.5, for any bounded sequence $b_k \in V_k$ there exists a bounded solution $v_k \in V_k$ of (2.153). By the Maizel' theorem (Theorem 2.1.1), the sequence B_k is hyperbolic on $\mathbb{Z}_-$ and $\mathbb{Z}_+$.

By the Pliss theorem (Theorem 2.1.2),

$$\mathscr{E}^s + \mathscr{E}^u = V_0, \tag{2.156}$$

where

$$\mathscr{E}^s = \{w_0 : w_{k+1} = B_k w_k, |w_k| \to 0, k \to \infty\}$$

and

$$\mathscr{E}^u = \{w_0 : w_{k+1} = B_k w_k, |w_k| \to 0, k \to -\infty\}.$$

Clearly, it follows from the hyperbolicity of the sequence B_k on $\mathbb{Z}_-$ and $\mathbb{Z}_+$ that the following equalities hold:

$$\mathscr{E}^s = \{w_0 : w_{k+1} = B_k w_k, \sup_{k \geq 0} |w_k| < \infty\}$$

and

$$\mathscr{E}^u = \{w_0 : w_{k+1} = B_k w_k, \sup_{k \leq 0} |w_k| < \infty\}.$$

We claim that

$$\mathscr{E}^s \subset E^s \text{ and } E^u \subset \mathscr{E}^u. \tag{2.157}$$

First we note that (2.157) implies the desired transversality of $W^s(p_0)$ and $W^u(p_0)$ at p_0.

Indeed, combining equality (2.156) with inclusions (2.157) and the trivial relations

$$E^s = V_0 \cap E^s + \{X(p_0)\} \text{ and } E^u = V_0 \cap E^u + \{X(p_0)\},$$

we conclude that

$$E^s + E^u = T_{p_0} M,$$

which gives us the transversality of $W^s(p_0)$ and $W^u(p_0)$ at p_0.

Thus, it remains to prove inclusions (2.157). We prove the first inclusion; for the second one, the proof is similar.

Case 1: The limit trajectory $x_0(t) = x_0$ is a rest point of X. In this case, the stable manifold of the rest point x_0 in the flow ϕ coincides with the stable manifold of the fixed point x_0 for the time-one diffeomorphism $f(x) = \phi(1, x)$.

It is clear that if p_k is a trajectory of f belonging to the stable manifold of x_0, then the tangent space to the stable manifold at p_0 is the subspace E^s of the initial values of bounded solutions of

$$v_{k+1} = A_k v_k, \quad k \geq 0. \tag{2.158}$$

Let us prove that $\mathscr{E}^s \subset E^s$. Fix an arbitrary sequence w_k such that $w_{k+1} = B_k w_k$ and $w_0 \in \mathscr{E}^s$. Consider the sequence

$$v_k = \lambda_k X(p_k)/|X(p_k)| + w_k,$$

where the λ_k satisfy the relations

$$\lambda_{k+1} = \frac{|X(p_{k+1})|}{|X(p_k)|}\lambda_k + \frac{X(p_{k+1})^*}{|X(p_{k+1})|}A_k w_k \tag{2.159}$$

(we denote by X^* the row-vector corresponding to the column-vector X) and $\lambda_0 = 0$. It is easy to see that the sequence v_k satisfies (2.158).

Since $x(t)$ is in the stable manifold of the hyperbolic rest point x_0, there exist positive constants K and α such that

$$\left|\frac{dx}{dt}(t)\right| \le K \exp(\alpha(t-s)) \left|\frac{dx}{dt}(s)\right|, \quad 0 \le s \le t.$$

This implies that

$$|X(p_k)| \le K \exp(\alpha(k-m)) \, |X(p_m)|, \quad 0 \le m \le k.$$

Thus, the scalar difference equation

$$\lambda_{k+1} = \frac{|X(p_{k+1})|}{|X(p_k)|}\lambda_k$$

is hyperbolic on $\mathbb{Z}_+$ and is, in fact, stable. Since the second term on the right in (2.159) is bounded as $k \to \infty$ (recall that we take $w_0 \in \mathscr{E}^s$), it follows that the λ_k are bounded for any choice of λ_0.

We conclude that v_k is a bounded solution of (2.158), and $v_0 = w_0 \in E^s$. Thus, we have shown that $\mathscr{E}^s \subset E^s$, which completes the proof in Case 1.

Case 2: The limit trajectory is in the set Σ (the chain recurrent set minus rest points). We know that the set Σ is hyperbolic. Our goal is to find the intersection of its stable manifold near $p_0 = x(0)$ with the cross-section at p_0 orthogonal to the vector field (in local coordinates generated by the exponential mapping). To do this, we discretize the problem and note that there exists a number $\sigma > 0$ such that a point p close to p_0 belongs to $W^s(p_0)$ if and only if the distances of the consecutive points of intersections of the positive semitrajectory of p to the points p_k do not exceed σ.

For suitably small $\mu > 0$ we find all the sequences of numbers t_k and vectors $z_k \in V_k$ (recall that V_k is the orthogonal complement to $\{X(p_k)\}$ at p_k) such that

$$|t_k - 1| \leq \mu, \ |z_k| \leq \mu, \ y_{k+1} = \phi(t_k, y_k), \quad k \geq 0,$$

where $y_k = p_k + z_k$.

Thus, we have to solve the equations

$$p_{k+1} = \phi(t_k, p_k + v_k), \quad k \geq 0,$$

for numbers t_k and vectors $z_k \in V_k$ such that $|t_k - 1| \leq \mu$ and $|z_k| \leq \mu$.

We reduce this problem to an equation in a Banach space. It was mentioned above that the sequence $\{B_k\}$ generating the difference equation

$$z_k = B_k z_k, \quad k \geq 0,$$

(where $B_k = P_{k+1}A_k$ and P_k is the orthogonal projection with range V_k) is hyperbolic on $\mathbb{Z}_+$. Denote by $Q_k : V_k \to V_k$ the corresponding projections to the stable subspaces and by $\mathscr{R}(Q_0)$ the range of Q_0 (note that $\mathscr{R}(Q_0) = \mathscr{E}^s$).

Fix a positive number μ_0 and denote by $\mathscr{V}$ the space of sequences

$$\mathscr{V} = \{z_k \in V_k : |z_k| \leq \mu_0, \ k \in \mathbb{Z}_+\}.$$

Let $l^\infty(\mathbb{Z}_+, \{\mathscr{M}_{k+1}\})$ be the space of sequences $\{\zeta_k \in \mathscr{M}_{k+1} : k \in \mathbb{Z}_+\}$ with the usual norm.

Define a C^1 function

$$G : [1 - \mu_0, 1 + \mu_0]^{\mathbb{Z}_+} \times \mathscr{V} \times \mathscr{R}(Q_0) \to l^\infty(\mathbb{Z}_+, \{\mathscr{M}_{k+1}\}) \times \mathscr{R}(Q_0)$$

by

$$G(t, z, \eta) = (\{p_{k+1} + z_{k+1} - \phi(t_k, p_k + z_k)\}, Q_0 z_0 - \eta).$$

This function is defined if μ_0 is small enough.

We want to solve the equation

$$G(t, z, \eta) = 0$$

for (t, z) as a function of η. It is clear that

$$G(1, 0, 0) = 0,$$

where the first argument of G is $\{1, 1, \dots\}$, the second argument is $\{0, 0, \dots\}$, and the right-hand side is $(\{0, 0, \dots\}, 0)$.

To apply the implicit function theorem, we must verify that the operator

$$T = \frac{\partial G}{\partial(t,z)}(1,0,0)$$

is invertible.

First we note that if $(s,w) \in l^\infty(\mathbb{Z}_+, \{\mathscr{M}_{k+1}\}) \times \mathscr{V}$, then

$$T(s,w) = (\{w_{k+1} - X(p_{k+1})s_k - A_k w_k\}, Q_0 w_0).$$

To show that T is invertible, we have to show that the equation

$$T(s,w) = (g,\eta)$$

has a unique solution for any $(g,\eta) \in l^\infty(\mathbb{Z}_+, \{\mathscr{M}_{k+1}\}) \times \mathscr{R}(Q_0)$. Thus, we have to solve the equation

$$w_{k+1} = A_k w_k + X(p_{k+1})s_k = g_k, \quad k \geq 0, \tag{2.160}$$

subject to the condition

$$Q_0 w_0 = \eta.$$

If we multiply Eq. (2.160) by $X(p_{k+1})^*$ and solve for s_k, we get the equalities

$$s_k = -\frac{X(p_{k+1})^*}{|X(p_{k+1})|^2}[A_k w_k + g_k], \quad k \geq 0,$$

and if we multiply Eq. (2.160) by P_{k+1}, we get the equalities

$$w_{k+1} = P_{k+1}A_k w_k + P_{k+1}g_k = B_k w_k + P_{k+1}g_k, \quad k \geq 0.$$

Now we know that the last equations have a unique bounded solution $w_k \in V_k$, $k \geq 0$, that satisfies $Q_0 w_0 = \eta$. Thus, T is invertible.

Hence, we can apply the implicit function theorem to show that there exists a $\mu > 0$ such that if $|\eta|$ is sufficiently small, then the equation $G(t,z,\eta) = 0$ has a unique solution $(t(\eta), z(\eta))$ such that $\|t-1\|_\infty \leq \mu$ and $\|z\|_\infty \leq \mu$. Moreover, $t(0) = 1$, $z(0) = 0$, and the functions $t(\eta)$ and $z(\eta)$ are of class C^1.

The points $p_0 + z_0(\eta)$ form a submanifold containing p_0 and contained in $W^s(p_0)$. Thus, the range of the derivative $z_0'(0)$ is contained in E^s.

Take an arbitrary vector $\xi \in \mathscr{E}^s$ and consider $\eta = \tau\xi$, $\tau \in \mathbb{R}$. Differentiating the equalities

$$p_{k+1} + z_{k+1}(\tau\xi) = \phi(t_k(\tau\xi), p_k + z_k(\tau\xi)), \quad k \geq 0,$$

and

$$Q_0(\tau\xi) = \tau\xi$$

with respect to τ at $\tau = 0$, we see that

$$s_k = \frac{\partial t_k}{\partial \eta}|_{\eta=0}\xi \text{ and } w_k = \frac{\partial z_k}{\partial \eta}|_{\eta=0}\xi \in V_k$$

are bounded sequences satisfying the equalities

$$w_{k+1} = A_k w_k + X(p_{k+1})s_k \text{ and } Q_0 w_0 = \xi.$$

Multiplying by P_{k+1}, we conclude that

$$w_{k+1} = B_k w_k \text{ and } Q_0 w_0 = \xi.$$

It follows that $w_0 \in \mathscr{E}^s = \mathscr{R}(Q_0)$. Then $w_0 = Q_0 w_0 = \xi$.

We have shown that the range of $z_0'(0)$ is exactly $\mathscr{E}^s$. Thus, $\mathscr{E}^s \subset E^s$.

Historical Remarks Theorem 2.7.1 was published by K. Palmer, the first author, and S. B. Tikhomirov in [57].

Chapter 3
C^1 Interiors of Sets of Systems with Various Shadowing Properties

In this chapter, we study the structure of C^1 interiors of some basic sets of dynamical systems having various shadowing properties. We give either complete proofs or schemes of proof of the following main results:

- The C^1 interior of the set of diffeomorphisms having the standard shadowing property is a subset of the set of structurally stable diffeomorphisms (Theorem 3.1.1); this result and Theorem 1.4.1 (a) imply that the C^1 interior of the set of diffeomorphisms having the standard shadowing property coincides with the set of structurally stable diffeomorphisms;
- the set $\mathrm{Int}^1(\mathrm{OrientSP}_F \setminus \mathscr{B})$ is a subset of the set of structurally stable vector fields (Theorem 3.3.1); similarly to the case of diffeomorphisms, this result and Theorem 1.4.1 (b) imply that the set $\mathrm{Int}^1(\mathrm{OrientSP}_F \setminus \mathscr{B})$ coincides with the set of structurally stable vector fields;
- the set $\mathrm{Int}^1(\mathrm{OrientSP}_F)$ contains vector fields that are not structurally stable (Theorem 3.4.1).

The structure of the chapter is as follows.

Section 3.1 is devoted to the proof of Theorem 3.1.1:

$$\mathrm{Int}^1(\mathrm{SSP}_D) \subset \mathscr{S}_D.$$

Our proof of Theorem 3.1.1 is based on reduction to Theorem 1.3.6 (2) (the C^1 interior of the set of Kupka–Smale diffeomorphisms coincides with the set of structurally stable diffeomorphisms).

We give a detailed proof of the inclusion

$$\mathrm{Int}^1(\mathrm{SSP}_D) \subset \mathrm{HP}_D$$

S.Yu. Pilyugin, K. Sakai, *Shadowing and Hyperbolicity*, Lecture Notes in Mathematics 2193, DOI 10.1007/978-3-319-65184-2_3

(thus, any periodic point of a diffeomorphism $f \in \text{Int}^1(\text{SSP}_D)$ is hyperbolic). Concerning the proof of transversality of stable and unstable manifolds of periodic points of a diffeomorphism $f \in \text{Int}^1(\text{SSP}_D)$, we refer the reader to Sect. 3.3 where a similar statement is proved in a more complicated case of flows on manifolds.

One of the necessary and sufficient conditions of structural stability of a diffeomorphism is Axiom A. In Sect. 3.2, we give an independent proof of the following statement, Theorem 3.2.1: If $f \in \text{Int}^1(\text{SSP}_D)$, then f satisfies Axiom A. Our proof uses neither Mañé's ergodic closing lemma [42] nor the techniques of creating homoclinic orbits developed in [44]. Instead, we refer to a sifting type lemma of Wen–Gan–Wen [109] influenced by Liao's work and apply it to Liao's closing lemma.

Sections 3.3 and 3.4 are devoted to the study of the C^1 interior of the set of vector fields having the oriented shadowing property. We introduce a special class $\mathscr{B}$ of vector fields having two rest points p and q for which there exists a trajectory of nontransverse intersection of the stable manifold $W^s(p)$ and $W^u(q)$. Of course, vector fields in $\mathscr{B}$ are not structurally stable.

In Sect. 3.3, we prove Theorem 3.3.1: The set

$$\text{Int}^1(\text{OrientSP}_F \setminus \mathscr{B})$$

is a subset of the set of structurally stable vector fields.

At the same time, we show in Sect. 3.4 that the set $\text{Int}^1(\text{OrientSP}_F)$ contains vector fields belonging to $\mathscr{B}$. The complete description of the corresponding example given in [69] is quite complicated, and we describe a "model" suggested in [100].

3.1 C^1 Interior of SSP_D

The main result of this section is the following theorem.

Theorem 3.1.1 *$Int^1(SSP_D) \subset \mathscr{S}_D$.*

It follows from Theorem 1.4.1 (a) that

$$\mathscr{S}_D \subset \text{LSP}_D \subset \text{SSP}_D.$$

Since the set of structurally stable diffeomorphisms is C^1-open,

$$\mathscr{S}_D = \text{Int}^1(\mathscr{S}_D) \subset \text{Int}^1(\text{SSP}_D).$$

Combining this with Theorem 3.1.1, we conclude that the C^1 interior of the set of diffeomorphisms having the standard shadowing property coincides with the set of structurally stable diffeomorphisms.

As was said at the beginning of this chapter, we reduce the proof of Theorem 3.1.1 to Theorem 1.3.6 (2). Thus, we have to show that

$$\mathrm{Int}^1\,(\mathrm{SSP}_D) \subset \mathrm{Int}^1\,(\mathrm{KS}_D).$$

Of course, for this purpose, it is enough to show that

$$\mathrm{Int}^1\,(\mathrm{SSP}_D) \subset \mathrm{KS}_D. \tag{3.1}$$

This means that we have to establish the inclusion

$$\mathrm{Int}^1(\mathrm{SSP}_D) \subset \mathrm{HP}_D \tag{3.2}$$

(i.e., every periodic point of a diffeomorphism in $\mathrm{Int}^1\,(\mathrm{SSP}_D)$ is hyperbolic) and to show that, for a diffeomorphism in $\mathrm{Int}^1\,(\mathrm{SSP}_D)$, stable and unstable manifolds of its periodic points arc transverse.

We prove inclusion (3.2) in Lemma 3.1.2.

We do not give here a proof of transversality of stable and unstable manifolds of periodic points of a diffeomorphism in $\mathrm{Int}^1\,(\mathrm{SSP}_D)$. Instead, we refer the reader to Sect. 3.3 of this book; in this section, a similar statement is proved for the case of vector fields (which is technically really more complicated). We advice the reader to "transfer" the proof of Sect. 3.3 to the case of diffeomorphisms.

We start with a lemma proved by Franks in [19]; this lemma plays an essential role in proofs of several theorems below.

If U is a domain in $\mathbb{R}^m$ with compact closure and $f, g : U \to \mathbb{R}^m$ are diffeomorphisms of U onto their images such that $f(U) = g(U) = V$, then we define $\rho_{1,U}(f, g)$ as the maximum of the following values:

$$\sup_{x\in U} \left|f(x) - g(x)\right|, \quad \sup_{x\in U} \left\|Df(x) - Dg(x)\right\|,$$

$$\sup_{y\in V} \left|f^{-1}(y) - g^{-1}(y)\right|, \quad \sup_{y\in V} \left\|Df^{-1}(y) - Dg^{-1}(y)\right\|$$

(this definition corresponds to our definition of the C^1 topology of $\mathrm{Diff}^1(M)$, see Sect. 1.3).

Lemma 3.1.1 *Let U be a domain in $\mathbb{R}^m$ with compact closure, where $m \geq 1$, and let $f : U \to \mathbb{R}^m$ be a C^1 diffeomorphism of U onto its image.*

Consider a finite set of different points $\{x_1, x_2, \dots, x_n\} \subset U$.

Then for any $\varepsilon > 0$, any neighborhood N of the set $\{x_1, x_2, \dots, x_n\}$, and any linear isomorphisms

$$L_i : \mathbb{R}^m \to \mathbb{R}^m$$

such that

$$\|L_i - Df(x_i)\|, \|L_i^{-1} - (Df(x_i))^{-1}\| \le \varepsilon/8, \quad 1 \le i \le n, \tag{3.3}$$

there exists a number $\delta > 0$ *and a* C^1 *diffeomorphism* $g : U \to \mathbb{R}^m$ *with* $f(U) = g(U)$ *and such that*

(a) $\rho_{1,U}(f, g) \le \varepsilon,$
(b) $g(x) = f(x), \quad x \in U \setminus N,$
and
(c) $g(x) = f(x_i) + L_i(x - x_i), \quad x \in N(\delta/4, x_i),\ 1 \le i \le n.$

Proof Standard reasoning shows that since U is a domain with compact closure, there exists a number $\varepsilon_0 > 0$ such that if g is a C^1 mapping of U such that $f(U) = g(U)$ and

$$\sup_{x \in U} |f(x) - g(x)|, \quad \sup_{x \in U} \|Df(x) - Dg(x)\| < \varepsilon_0,$$

then g is a diffeomorphism of U onto $g(U)$.

For a positive $\delta > 0$, let

$$B_\delta(x_i) = \{y \in U : |y - x_i| \le \delta\}, \quad 1 \le i \le n.$$

We assume that δ is small enough, so that the sets $B_\delta(x_i)$ with different i do not intersect. In what follows, we reduce δ if necessary.

Choose a C^∞ real-valued function $\sigma : \mathbb{R} \to \mathbb{R}$ such that $0 \le \sigma(x) \le 1$,

$$\sigma(x) = \begin{cases} 0 \text{ if } |x| \ge \delta, \\ 1 \text{ if } |x| \le \delta/4, \end{cases}$$

and $0 \le |\sigma'(x)| < 2/\delta$ for all x.

Let $\rho : \bigcup_{i=1}^n B_\delta(x_i) \to \mathbb{R}$ be defined by

$$\rho(y) = \sigma(|y - x_i|), \quad y \in B_\delta(x_i),\ 1 \le i \le n.$$

Fix $\varepsilon \in (0, \varepsilon_0)$ and take $0 < \delta < \min(1, \varepsilon)$ so small that

$$\bigcup_{i=1}^n B_\delta(x_i) \subset N, \tag{3.4}$$

$$|f(x_i) + L_i(y - x_i) - f(y)| \le \frac{\varepsilon}{4}|y - x_i|, \tag{3.5}$$

and

$$|L_i v - Df(y)v| \le \frac{\varepsilon}{4}|v|, \quad v \in \mathbb{R}^m, \tag{3.6}$$

for $y \in B_\delta(x_i)$, $1 \le i \le n$ (clearly, this is possible due to estimates (3.3)).

Define a mapping $g : U \to \mathbb{R}^m$ by

$$g(y) = \begin{cases} f(y) & \text{if } y \notin \bigcup_{i=1}^n B_\delta(x_i), \\ \rho(y)(f(x_i) + L_i(y - x_i)) + (1 - \rho(y))f(y) & \text{if } y \in \bigcup_{i=1}^n B_\delta(x_i). \end{cases}$$

It is easy to see that if $y \in \bigcup_{i=1}^n B_\delta(x_i)$, then

$$|f(y) - g(y)| = |\rho(y)(f(x_i) + L_i(y - x_i)) - \rho(y)f(y)| =$$

$$= \rho(y)|f(x_i) + L_i(y - x_i) - f(y)| \le 1 \cdot \frac{\varepsilon}{4} \cdot \delta < \varepsilon.$$

Let us estimate the differences of the derivatives. If $y \in B_\delta(x_i)$ and $v \in \mathbb{R}^m$, then

$$Dg(y)v = \rho(y)L_i v + \langle D\rho(y), v\rangle(f(x_i) + L_i(y - x_i)) +$$

$$+(1 - \rho(y))Df(y)v - \langle D\rho(y), v\rangle f(y),$$

where

$$\langle D\rho(y), v\rangle = \sum_{j=1}^m \frac{\partial \rho}{\partial y_j}(y)v_j.$$

Thus,

$$|Df(y)v - Dg(y)v| =$$

$$= |\rho(y)L_i v - \rho(y)Df(y)v + \langle D\rho(y), v\rangle(f(x_i) + L_i(y - x_i)) - \langle D\rho(y), v\rangle f(y)| \le$$

$$\le \rho(y)|L_i v - Df(y)v| + |\langle D\rho(y), v\rangle||f(x_i) + L_i(y - x_i) - f(y)|.$$

It is clear that if $|y - x_i| > \delta$, then $\rho(y) = 0$, and if $|y - x_i| \le \delta$, then, by the choice of δ (see (3.6)),

$$\rho(y) \cdot |L_i v - Df(y)v| \le |L_i v - Df(y)v| \le \frac{\varepsilon}{4}|v|.$$

If $|y - x_i| > \delta$, then $D\rho(y) = 0$ (since $\rho(y) = 0$ for $|y - x_i| > \delta$). If $|y - x_i| \le \delta$, then $|D\rho(y)| < 2/\delta$ and

$$|f(x_i) + L_i(y - x_i) - f(y)| \le \frac{\varepsilon}{4}|y - x_i|$$

by the choice of δ (see (3.5)) and the definition of ρ. Thus,

$$|\langle D\rho(y), v\rangle||f(x_i) + L_i(y - x_i) - f(y)| \leq$$

$$\leq \frac{2}{\delta} \cdot \frac{\varepsilon}{4}|y - x_i||v| \leq \frac{2}{\delta} \cdot \frac{\varepsilon}{4}\delta|v| = \frac{\varepsilon}{2}|v|.$$

Hence,

$$|Df(y)v - Dg(y)v| \leq \frac{\varepsilon}{4}|v| + \frac{\varepsilon}{2}|v| \leq \varepsilon|v|.$$

It follows from the choice of $\varepsilon < \varepsilon_0$ that g is a diffeomorphism of U onto $g(U) = f(U)$.

Now a similar reasoning can be applied to estimate the values

$$|f^1(y) - g^{-1}(y)| \text{ and } \|Df^1(y) - Dg^{-1}(y)\|$$

(reducing δ, if necessary).

Inclusion (3.4) implies that g and f coincide outside N. The lemma is proved. □

Lemma 3.1.2 *Inclusion (3.2) holds.*

Proof Let us consider the case of an m-dimensional manifold M with $m \geq 1$. To get a contradiction, assume that there exists a diffeomorphism $f \in \text{Int}^1(\text{SSP}_D(M)) \setminus \text{HP}_D(M)$.

Then f has a nonhyperbolic periodic point p of period $\pi(p)$.

Take a C^1 neighborhood $\mathscr{U}(f)$ of f lying in $\text{SSP}_D(M)$.

To simplify presentation, we assume that $\pi(p) = 1$ (the case of a periodic point of minimal period $\pi(p) > 1$ is considered similarly). Moreover, since the argument is local, we assume further that f is defined on an open set of $\mathbb{R}^m$.

By the Franks lemma, it is possible to find a diffeomorphism $g \in \mathscr{U}(f)$ with the following properties:

- p is a fixed point of g,
- g is linear in a neighborhood of p.

Indeed, let us introduce local coordinates $x \in \mathbb{R}^m$ near p such that p is the origin. Then, by the Franks lemma, for any $r > 0$ there exists a diffeomorphism f_r such that

- $f_r(x) = f(x)$ for $x \notin N(4r, p)$,
- $f_r(x) = Df(p)x$ for $x \in N(r, p)$.

Note that f_r converges to f with respect to the C^1 topology as $r \to 0$. Fix $r_0 > 0$ such that $f_{r_0} \in \mathscr{U}(f)$ and write g instead of f_{r_0}.

Since the point $p = 0$ is not hyperbolic, the matrix $Dg(p)$ has an eigenvalue λ with $|\lambda| = 1$. To simplify presentation, we assume that $\lambda = 1$ (the proof in the general case can be found in [87]).

Applying a C^1-small perturbation of g (so that the perturbed g still is in $\mathscr{U}(f)$) and preserving the notation g for the perturbed diffeomorphism, we may assume further that $Dg(p)$ has an eigenvalue equal to 1, p is the origin with respect to some local coordinates $x = (x_1, \dots, x_m)$, and g maps a point $x = (x_1, y) \in N(r_0, p)$, where $y = (x_2, \dots, x_m)$, to the point (x_1, By), where B is a hyperbolic matrix.

In this case, the segment

$$\mathscr{I} = \{(x_1, 0, \dots, 0) : 0 \le |x_1| \le r_0\}$$

consists of fixed points of g.

Since it was assumed that $g \in SSP_D(M)$, for $\varepsilon = r_0/2$ there is the corresponding $0 < d < \varepsilon$ from the definition of the standard shadowing property. Take a natural number l such that the sequence

$$\xi = \{x_k : k \in \mathbb{Z}\} \subset \mathscr{I},$$

where

$$x_k = \begin{cases} 0 & \text{for } k < 0; \\ \left(\frac{r_0 k}{2l}, 0, \dots, 0\right) & \text{for } 0 \le k \le l; \\ (r_0/2, 0, \dots, 0) & \text{for } k > l, \end{cases}$$

is a d-pseudotrajectory of g.

Let $x \in N(\varepsilon, x_0)$ be a point such that

$$|g^k(x) - x_k| < \epsilon \quad \text{for} \quad k \in \mathbb{Z}.$$

Since the matrix B is hyperbolic, for any point (x_1, y) with $y \neq 0$, its g-trajectory leaves the set $N(r_0, p)$. Hence, if

$$|g^k(x) - x_k| < \varepsilon, \quad k \in \mathbb{Z},$$

then $x = (b, 0, \dots, 0)$. Since

$$g(x) = g(b, 0, \dots, 0) = (b, 0, \dots, 0),$$

we see that $|b| < r_0/2$, and then $|b - r_0| < r_0/2$. The obtained contradiction proves our lemma. □

Historical Remarks One of the first results concerning C^1 interiors of sets of diffeomorphisms with properties similar to shadowing was proved by K. Moriyasu in [47].

Let us denote by TS_D the set of topologically stable diffeomorphisms. Recall that a diffeomorphism f of a smooth manifold M is called *topologically stable* if for any

$\varepsilon > 0$ there is a $d > 0$ such that for any homeomorphism g satisfying the inequality $\rho_0(f, g) < d$, there exists a continuous map h mapping M onto M and such that $\rho_0(h, id) < \varepsilon$ and $f \circ h = h \circ g$ (see [104]).

It is known that every topologically stable diffeomorphism has the standard shadowing property (see [46, 105]); thus, $\text{SSP}_D \subset \text{TS}_D$. In addition, every expansive diffeomorphism in SSP_D is in TS_D (see [64] for details).

K. Moriyasu proved in [47] that any diffeomorphism in $\text{Int}^1(\text{TS}_D)$ satisfies Axiom A. In fact, the paper [47] contains the proof of inclusion (3.2) (see Proposition 1 in [47]).

Theorem 3.1.1 was proved by the second author in [87].

Later, a more general result (in which the set SSP_D was replaced by a larger set OSP_D) was obtained by the first author, A. A. Rodionova, and the second author in [65] (the method of proving transversality of the stable and unstable manifolds of periodic points used in [65] was later applied in the case of vector fields [69]; see Sect. 3.3 of this book).

In [88], the second author introduced the notion of C^0 transversality and showed that for two-dimensional Axiom A diffeomorphisms, C^0 transversality of one-dimensional stable and unstable manifolds is equivalent to shadowing. Later, the authors related C^0 transversality to inverse shadowing in two-dimensional Axiom A diffeomorphisms [66].

Let us mention here one more result of that type related to shadowing properties. Let f be a homeomorphism of a metric space (M, dist). We say that f has the *weak shadowing property* if for any $\varepsilon > 0$ there exists $d > 0$ such that for any d-pseudotrajectory ξ of f there is a point $p \in M$ such that

$$\xi \subset N(\varepsilon, O(p, f)).$$

Denote by WSP_D the set of diffeomorphisms having the weak shadowing property.

It was shown by the second author in [89] that if M is a smooth two-dimensional manifold, then

$$\text{Int}^1(\text{WSP}_D(M)) \subset \Omega\mathcal{S}_D(M).$$

Let us note that the above inclusion is strict; it was shown by O. B. Plamenevskaya in [72] that there exist Ω-stable diffeomorphisms of the two-dimensional torus that do not have the weak shadowing property.

Let us also note that the result of [89] cannot be generalized to higher dimensions. R. Mañé constructed in [40] an example of a C^1-open subset $\mathcal{T}$ of the space of diffeomorphisms of the three-dimensional torus such that

- any diffeomorphism $f \in \mathcal{T}$ has a dense orbit (thus, any $f \in \mathcal{T}$ is in $\text{Int}^1(\text{WSP}_D)$);
- any diffeomorphism $f \in \mathcal{T}$ is not Anosov (and hence, it is not Ω-stable).

3.2 Diffeomorphisms in Int^1 (SSP$_D$) Satisfy Axiom A

As was said at the beginning of this chapter, in this section we prove the following statement.

Theorem 3.2.1 *If $f \in$ Int^1 (SSP$_D$), then f satisfies Axiom A.*

Remark 3.2.1

1. To get an independent proof of Theorem 3.1.1 using Theorem 3.2.1, one has to show that if a diffeomorphism $f \in$ Int^1 (SSP$_D$) satisfies Axiom A, then f also satisfies the strong transversality condition.

   This can be done by applying the following scheme. Assuming that the stable manifold $W^s(p)$ and the unstable manifold $W^u(q)$ for two points $p, q \in \Omega(f)$ have a point r of nontransverse intersection, one can approximate r by points of intersection of periodic points of f and then, perturbing f, to get a point of nontransverse intersection of periodic points of a diffeomorphism $g \in$ Int^1 (SSP$_D$). After that, one can apply the techniques described in Sect. 3.3 to get a contradiction. We leave details to the reader.
2. Of course, it has shown by Mañé and Hayashi [25, 42, 45] that a diffeomorphism $f \in$ Int^1 (HP$_D$) satisfies Axiom A, but we give a simpler proof of this result under the assumption that $f \in$ Int^1 (SSP$_D$); this proof uses neither Mañé's ergodic closing lemma [42] nor the techniques creating homoclinic orbits developed in [44].

Let the phase space be a ν-dimensional manifold M.

Denote, as above, by Per(f) the set of periodic points of a diffeomorphism $f : M \to M$. Let $\pi(p)$ be the minimal period of a periodic point $p \in$ Per(f).

It is proved in [40] that if $f \in$ Int^1 (SSP$_D(M)$, then $\Omega(f) =$ Cl(Per(f)).

Denote by $P_j(f)$, $0 \le j \le \nu$, the set of hyperbolic periodic points of f whose index (the dimension of the stable manifold) is equal to j. Let Λ_j be the closure of the set $P_j(f)$.

It has shown by Pliss [73] that the sets of sinks, $P_\nu(f)$, and of sources, $P_0(f)$, of a diffeomorphism $f \in$ Int^1(SSP$_D(M)$) are finite sets (another proof can be found in [36]).

The following lemma is a "globalized" variant of Frank's lemma (Lemma 3.1.1) for C^1 diffeomorphisms of a smooth closed manifold using exponential mappings.

Lemma 3.2.1 *Let $f \in$ Diff$^1(M)$ and let $\mathscr{U}(f)$ be a neighborhood of f.*

Then there exists a number $\delta_0 > 0$ and a neighborhood $\mathscr{V}(f) \subset \mathscr{U}(f)$ such that for any $g \in \mathscr{V}(f)$, any finite set $\{x_1, x_2, \dots, x_m\}$ consisting of pairwise different points, any neighborhood U of the set $\{x_1, x_2, \dots, x_m\}$, and any linear isomorphisms $L_i : T_{x_i}M \to T_{g(x_i)}M$ such that

$$\|L_i - Dg(x_i)\|, \|L_i^{-1} - Dg^{-1}(x_i)\| \le \delta_0, \quad 1 \le i \le m,$$

there exist $\varepsilon_0 > 0$ *and* $\tilde{g} \in \mathcal{U}(f)$ *such that*

(*a*) $\tilde{g}(x) = g(x)$ *if* $x \in M \setminus U$, *and*
(*b*) $\tilde{g}(x) = \exp_{g(x_i)} \circ L_i \circ \exp_{x_i}^{-1}(x)$ *if* $x \in B_{\varepsilon_0}(x_i)$ *for all* $1 \le i \le m$.

Note that assertion (*b*) implies that $\tilde{g}(x) = g(x)$ if $x \in \{x_1, x_2, \dots, x_m\}$ and that $D_{x_i}\tilde{g} = L_i$ for all $1 \le i \le m$.

In what follows, we assume that $f \in \text{Int}^1(\text{SSP}_D)$; hence, by Lemma 3.1.1, $f \in \text{Int}^1(\text{HP}_D)$.

Thus, there exists a neighborhood $\mathcal{U}(f)$ of f in Diff$^1(M)$ such that every periodic point $p \in \text{Per}(g)$ is hyperbolic for any $g \in \mathcal{U}(f)$.

Then there exists a C^1 neighborhood $\mathcal{V}(f)$ of f such that the family of periodic sequences of linear isomorphisms of tangent spaces of M generated by the differentials Dg of diffeomorphisms $g \in \mathcal{V}(f)$ along hyperbolic periodic orbits of points $q \in \text{Per}(g)$ is uniformly hyperbolic (see [42]).

To be exact, this means that there exists $\varepsilon > 0$ and a neighborhood $\mathcal{V}(f)$ of f such that for any $g \in \mathcal{V}(f)$, any $q \in \text{Per}(g)$, and any sequence of linear maps

$$L_i : T_{g^i(q)}M \to T_{g^{i+1}(q)}M$$

with

$$\left\| L_i - Dg\left(g^i(q)\right) \right\| < \varepsilon, \quad i = 1, \dots, \pi(q) - 1,$$

$\prod_{i=0}^{\pi(q)-1} L_i$ is hyperbolic (here $\varepsilon > 0$ and $\mathcal{V}(f)$ correspond to $\mathcal{U}(f)$) according to Lemma 3.2.1.

The following result was proved by Mañé [42, Proposition II.1]. Denote by $E^s(q)(f)$ and $E^u(q)(f)$ the stable and unstable spaces of the hyperbolic structure at a point q of a hyperbolic periodic orbit of f, respectively.

Proposition 3.2.1 *Let* $f \in \text{Int}^1(HP_D)$.

In the above notation, there are constants $C > 0$, $m > 0$, *and* $0 < \lambda < 1$ *such that:*

(*a*) *if* $g \in \mathcal{V}(f)$, $q \in Per(g)$, *and* $\pi(q) \ge m$, *then*

$$\prod_{i=0}^{k-1} \left\| Dg^m_{|E^s(g^{im}(q))(g)} \right\| \le C\lambda^k \ \text{and} \ \prod_{i=0}^{k-1} \left\| Dg^{-m}_{|E^u(g^{-im}(q))(g)} \right\| \le C\lambda^k,$$

where $k = [\pi(q)/m]$.

(*b*) *For any* $g \in \mathcal{V}(f)$ *and* $0 \le j \le \nu$, *the set* $\Lambda_j(g) = Cl(P_j(g))$ *admits a dominated splitting (see Definition 1.3.12)*

$$T_{\Lambda_j(g)}M = E(g) \oplus F(g)$$

with $\dim E(g) = j$, *i.e.*,

$$\left\| Dg^m_{|E(x)(g)} \right\| \cdot \left\| Dg^{-m}_{|F(g^m(x))(g)} \right\| \leq \lambda$$

for all $x \in Cl(P_j(g))$ *(note that* $E(x)(g) = E^s(x)(g)$ *and* $F(x)(g) = E^u(x)(g)$ *if* $x \in P_j(g)$*).*

It is easy to see that the above proposition can be restated in the following way.

Proposition 3.2.2 *In the notation and assumptions of Proposition 3.2.1, there exist constants* $m > 0$, $0 < \lambda < 1$, *and* $L > 0$ *such that:*

(a) If $g \in \mathcal{V}(f)$, $q \in Per(g)$, *and* $\pi(q) \geq L$, *then*

$$\prod_{i=0}^{\pi(q)-1} \left\| Dg^m_{|E^s(g^{im}(q))(g)} \right\| < \lambda^{\pi(q)} \text{ and } \prod_{i=0}^{\pi(q)-1} \left\| Dg^{-m}_{|E^u(g^{-im}(q))(g)} \right\| < \lambda^{\pi(q)}.$$

(b) For any $g \in \mathcal{V}(f)$ *and* $0 \leq j \leq \nu$, *the set* $\Lambda_j(g)$ *admits a dominated splitting* $T_{\Lambda_j(g)}M = E(g) \oplus F(g)$ *with* $\dim E(g) = j$ *such that*

$$\left\| Dg^m_{|E(x)(g)} \right\| \cdot \left\| Dg^{-m}_{|F(g^m(x))(g)} \right\| < \lambda^2$$

for any $x \in \Lambda_j(g)$ *(note that* $E(x)(g) = E^s(x)(g)$ *and* $F(x)(g) = E^u(x)(g)$ *if* $x \in P_j(g)$*).*

In what follows, we need two technical lemmas (Lemmas 3.2.2 and 3.2.3). Denote by Λ a set $\Lambda_j = \mathrm{Cl}(P_j(f))$, where $0 \leq j \leq \nu$.

Lemma 3.2.2 deals with extension of the dominated splitting to a small neighborhood of Λ in M. Assume that Λ admits a dominated splitting $T_\Lambda M = E \oplus F$ for which there exist constants $m > 0$ and $0 < \lambda < 1$ such that

$$\left\| Df^m_{|E(x)} \right\| \cdot \left\| Df^{-m}_{|F(f^m(x))} \right\| \leq \lambda$$

for all $x \in \Lambda$. To simplify notation, denote f^m by f.

It is known (see [27]) that there exists a neighborhood U of Λ, a constant $\hat{\lambda} > 0$, $\lambda < \hat{\lambda} < 1$, and a continuous splitting $T_U M = \hat{E} \oplus \hat{F}$ with $\dim \hat{E} = \dim E$ such that

- $\hat{E}|_\Lambda = E$ and $\hat{F}|_\Lambda = F$;
- $Df(x)\hat{E}(x) = \hat{E}(f(x))$ if $x \in U \cap f^{-1}(U)$;
- $Df^{-1}(x)\hat{F}(x) = \hat{F}(f^{-1}(x))$ if $x \in U \cap f(U)$;
- $\left\| Df^k_{|\hat{E}(x)} \right\| \cdot \left\| Df^{-k}_{|\hat{F}(f^k(x))} \right\| < \hat{\lambda}^k$ if $x \in \bigcap_{i=-k}^{k} f^i(U)$ for $k \geq 0$.

The continuity of the differential Df implies the following statement (in which we have to shrink the neighborhood U of Λ if necessary).

Lemma 3.2.2 *In the above notation and assumptions of Proposition 3.2.1, there exists a Df-invariant continuous splitting* $T_{\Lambda_f(U)}M = \hat{E} \oplus \hat{F}$ *with* $\dim \hat{E} = \dim E$ *and* $0 < \hat{\lambda} < 1$ *such that*

- $\hat{E}_{|\Lambda} = E$ *and* $\hat{F}_{|\Lambda} = F$;
- $\left\| Df^k_{|\hat{E}(x)} \right\| \cdot \left\| Df^{-k}_{|\hat{F}(f^k(x))} \right\| < \hat{\lambda}^k$ *for any* $x \in \Lambda_f(U)$ *and* $k \geq 0$;
- *for any* $\varepsilon > 0$ *there exists* $\delta > 0$ *such that if* $x \in \Lambda_f(U)$, $y \in \Lambda$, *and* $dist(x, y) < \delta$, *then*

$$\left| \log \left\| Df_{|\hat{E}(x)} \right\| - \log \left\| Df_{|E(y)} \right\| \right| < \varepsilon$$

and

$$\left| \log \left\| Df^{-1}_{|\hat{F}(x)} \right\| - \log \left\| Df^{-1}_{|F(y)} \right\| \right| < \varepsilon.$$

In the statement above,

$$\Lambda_f(U) = \bigcap_{n \in \mathbb{Z}} f^n(U).$$

The second technical lemma (Lemma 3.2.3) is a variant of the so-called sifting lemma first proved by Liao (see [36]). The statement which we prove belongs to Wen–Gan–Wen [109].

Let $T_{\Lambda_f(U)}M = \hat{E} \oplus \hat{F}$ be as in Lemma 3.2.2 and let $0 < \lambda < 1$.

An orbit string

$$\{x, n\} = \{x, f(x), \dots, f^n(x)\} \subset \Lambda_f(U)$$

is called a *λ-quasi-hyperbolic string* with respect to the splitting $\hat{E} \oplus \hat{F}$ if the following conditions are satisfied:

(1)

$$\prod_{i=0}^{k-1} \left\| Df_{|\hat{E}(f^i(x))} \right\| \leq \lambda^k \text{ for } k = 1, 2, \dots, n;$$

(2)

$$\prod_{i=k-1}^{n-1} m\left(Df_{|\hat{F}(f^i(x))} \right) \geq \lambda^{k-n-1} \text{ for } k = 1, 2, \dots, n;$$

(3)

$$\left\| Df_{|\hat{E}(f^i(x))} \right\| / m\left(Df_{|\hat{F}(f^i(x))} \right) \leq \lambda^2 \text{ for every } i = 0, 1, \dots, n-1.$$

Here $m(A)$ is the minimum norm of a linear map A, i.e.,

$$m(A) = \inf_{\|v\|=1} \|Av\|.$$

Lemma 3.2.3 (Sifting Lemma, [36, 107, 109]) *Let $\{a_i\}_{i=0}^{\infty}$ be an infinite sequence for which there exists a constant K such that $|a_i| < K$. Assume that*

$$\limsup_{n\to\infty} \frac{1}{n}\sum_{i=0}^{n-1} a_i = \xi \text{ and } \liminf_{n\to\infty} \frac{1}{n}\sum_{i=0}^{n-1} a_i = \xi',$$

where $\xi' < \xi$. Then for any ξ_1 and ξ_2 with $\xi_1 < \xi < \xi_2$ there is an infinite sequence $\{m_i\}_{i=1}^{\infty} \subset \mathbb{N}$ such that

$$\frac{1}{k}\sum_{j=m_i}^{m_i+k-1} a_j \le \xi_2 \text{ and } \frac{1}{k}\sum_{j=m_{i+1}-k}^{m_{i+1}-1} a_j \ge \xi_1$$

for every $i = 1, 2, \ldots$ and every $k = 1, \ldots, m_i + 1 - m_i$.

Proof Let $S(n) = \sum_{i=0}^{n-1} a_i$.

Fix a small $\varepsilon > 0$ with

$$\frac{\xi - \xi'}{2} > \varepsilon.$$

(We determine ε at the end of the proof.)

Choose a large enough $N \in \mathbb{N}$ such that

$$\frac{1}{n} S(n) < \xi + \varepsilon$$

for any $n > N$.

By our assumption, the upper and lower limits are different; hence, there is an infinite sequence

$$N < n_1 < n_1' < n_2 < n_2' < n_3 < n_3' < \ldots$$

such that

$$\frac{1}{n_i} S(n_i) < \xi' + \varepsilon < \xi - \varepsilon < \frac{1}{n_i'} S(n_i')$$

for every $i = 1, 2, \ldots$.

Take an integer $n_i < m_i < n_{i+1}$ such that

$$\frac{S(k) - S(m_i)}{k - m_i} \leq \xi - \varepsilon$$

for every $k = m_i + 1, m_+2 \ldots, n_{i+1}$ and

$$\frac{S(m_i) - S(k)}{m_i - k} \geq \xi - \varepsilon$$

for every $k = n_i, n_i + 1, \ldots, m_i - 1$.

This is a crucial point of the proof. Roughly speaking, m_i is the index at which $S(k) - S(n_i) - (k - n_i)(\xi - \varepsilon)$ attains maximum when k runs over the set $n_i + 1, n_i + 2, \ldots, n_{i+1}$ (Fig. 3.1).

Claim

$$n_{i+1} - m_i > \frac{\xi - \xi' - 2\varepsilon}{K + \xi' + \varepsilon} m_i \quad \text{and} \quad m_i - n_i > \frac{\xi - \xi' - 2\varepsilon}{K - \xi' - \varepsilon} m_i.$$

Proof (of the claim) By the choice of m_i, it is easy to see that

$$S(m_i) - S(n_i') \geq (m_i - n_i')(\xi - \varepsilon).$$

Hence,

$$S(m_i) \geq m_i(\xi - \varepsilon).$$

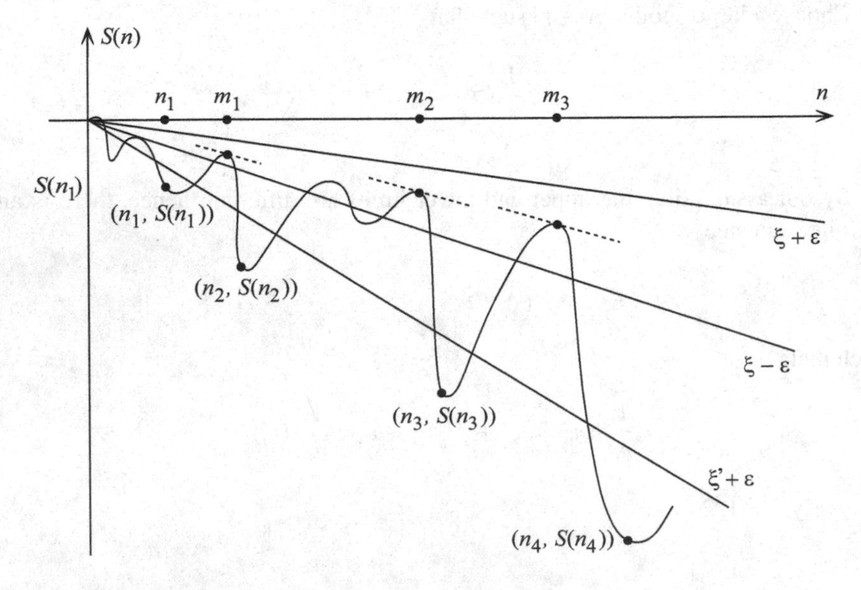

Fig. 3.1 The choice of m_i

Since $|a_i| < K$, we get the inequalities

$$n_{i+1}(\xi' + \varepsilon) > S(n_i + 1) > S(m_i) - K(n_{i+1} - m_i) \geq m_i(\xi - \varepsilon) - K(n_{i+1} - m_i)$$

and

$$n_i(\xi' + \varepsilon) + K(m_i - n_i) > S(n_i) + K(m_i - n_i) > S(m_i) \geq m_i(\xi - \varepsilon).$$

Hence,

$$K(n_{i+1} - m_i) > (\xi - \varepsilon)m_i - (\xi + \varepsilon)n_{i+1} =$$

$$= (\xi - \xi' - 2\varepsilon)m_i + (\xi' + \varepsilon)(m_i - n_{i+1})$$

and

$$K(m_i - n_i) > m_i(\xi - \varepsilon) - n_i(\xi' + \varepsilon) =$$

$$= m_i(\xi - \xi' - 2\varepsilon) + (\xi' + \varepsilon)(m_i - n_i).$$

Therefore,

$$n_{i+1} - m_i > \frac{\xi - \xi' - 2\varepsilon}{K + \xi' + \varepsilon} m_i \text{ and } m_i - n_i > \frac{\xi - \xi' - 2\varepsilon}{K - \xi' - \varepsilon} m_i.$$

Thus, the claim is proved. □

Let us pass to the proof of Lemma 3.2.3.

It is obvious that for $k = 1, 2, \ldots, n_{i+1} - m_i$,

$$\frac{1}{k}\left(S(m_i + k) - S(m_i)\right) \leq \xi - \varepsilon.$$

For $k = n_{i+1} - m_i + 1, \ldots, m_{i+1} - m_i$,

$$\frac{1}{k}\left(S(m_i + k) - S(m_i)\right) < \frac{1}{k}\left((m_i + k)(\xi + \varepsilon) - m_i(\xi - \varepsilon)\right) =$$

$$= \xi + \varepsilon + 2\varepsilon \frac{m_i}{k} < \xi + \left(1 + 2\frac{K + \xi' + \varepsilon}{\xi - \xi' - 2\epsilon}\right)\varepsilon.$$

Note that in the third inequality we have used the above claim.

Similarly, for $k = 1, 2, \ldots, m_i - n_i$,

$$\frac{1}{k}\left(S(m_i) - S(m_i - k)\right) \geq \xi - \varepsilon,$$

and for $k = m_i - n_i + 1, \dots, m_i - m_{i-1}$,

$$\frac{1}{k}(S(m_i) - S(m_i - k)) > \frac{1}{k}(m_i(\xi - \varepsilon) - (m_i - k)(\xi + \varepsilon) =$$

$$= \xi + \varepsilon - 2\varepsilon\frac{m_i}{k} > \xi + \left(1 - 2\frac{K - \xi' - \varepsilon}{\xi - \xi' - 2\varepsilon}\right)\varepsilon.$$

Now choose ε small enough so that

$$\xi + \left(1 + 2\frac{K + \xi' + \varepsilon}{\xi - \xi' - 2\varepsilon}\right)\varepsilon < \xi_2$$

and

$$\min\left\{\xi - \varepsilon, \xi + \left(1 - 2\frac{K - \xi' - \varepsilon}{\xi - \xi' - 2\varepsilon}\right)\varepsilon\right\} > \xi_1.$$

This proves Lemma 3.2.3. □

A proof of the following lemma (in fact of its generalized version) is given at the end of this section (see Lemma 3.2.5).

Lemma 3.2.4 (Liao's Closing Lemma [36]) *Let $T_{\Lambda_f(U)}M = \hat{E} \oplus \hat{F}$ be a continuous Df-invariant splitting. For any $0 < \lambda < 1$ and any $\varepsilon > 0$ there is $\delta > 0$ such that for any λ-quasi-hyperbolic string $\{x, n\}$ of f in $\Lambda_f(U)$ with dist $(f^n(x), x) < \delta$, there is a periodic point $p \in M$ of f such that $f^n(p) = p$ and dist $\left(f^i(p), f^i(x)\right) \le \varepsilon$ for all $0 \le i \le n - 1$.*

In the following proposition, to simplify notation, we denote $\Lambda_f(U)$, $\hat{E} \oplus \hat{F}$, and $\hat{\lambda}$ by Λ, $E \oplus F$, and λ, respectively. The next proposition is proved by applying Lemmas 3.2.3 and 3.2.4.

Proposition 3.2.3 *Let Λ be a compact f-invariant set, let $0 < \lambda < 1$ be given, and assume that there is a continuous Df-invariant splitting $T_\Lambda M = E \oplus F$ such that*

$$\left\|Df_{|E(x)}\right\| \cdot \left\|Df^{-1}_{|F(f(x))}\right\| < \lambda^2$$

for any $x \in \Lambda$.

Assume that there exists a point $y \in \Lambda$ such that

$$\log \lambda < \log \lambda_1 = \limsup_{n\to\infty} \frac{1}{n}\sum_{i=0}^{n-1} \log \left\|Df_{|E(f^i(y))}\right\| < 0$$

and

$$\liminf_{n\to\infty} \frac{1}{n}\sum_{i=0}^{n-1} \log \left\|Df_{|E(f^i(y))}\right\| < \log \lambda_1.$$

Then for any λ_2 and λ_3 with $\lambda < \lambda_2 < \lambda_1 < \lambda_3 < 1$ and any neighborhood W of Λ there is a hyperbolic periodic point q of index $\dim E$ *such that $O(q,f) \subset W$,*

$$\prod_{i=0}^{k-1} \left\| Df_{|E^s(f^i(q))} \right\| \le \lambda_3^k, \text{ and } \prod_{i=k-1}^{\pi(q)-1} \left\| Df_{|E^s(f^i(q))} \right\| > \lambda_2^{\pi(q)-k+1}$$

for all $k = 1, 2, \ldots, \pi(q)$.

Furthermore, q can be chosen so that the period $\pi(q)$ is arbitrarily large.

Our Theorem 3.2.1 follows from the next proposition (this kind of result was first obtained in [109]).

Proposition 3.2.4 *Let Λ be a compact f-invariant set, and let $0 < \lambda < 1$ and $L > 1$ be given. Assume that f has the following properties* (P.1)–(P.4).

(P.1) *There is a homogeneous Df-invariant splitting $T_\Lambda M = E \oplus F$ such that*

$$\left\| Df_{|E(x)} \right\| \cdot \left\| Df^{-1}_{|F(f(x))} \right\| < \lambda^2$$

for any $x \in \Lambda$.

(P.2) *There is a compact neighborhood U of Λ such that if $q \in \Lambda_f(U) \cap Per(f)$ and $\pi(q) \ge L$, then*

$$\prod_{i=0}^{\pi(q)-1} \left\| Df_{|E^s(f^i(q))} \right\| < \lambda^{\pi(q)} \text{ and } \prod_{i=0}^{\pi(q)-1} \left\| Df^{-1}_{|E^u(f^{-i}(q))} \right\| < \lambda^{\pi(q)}.$$

(P.3) $\Lambda = \overline{P_{\dim E}(f)}$.

(P.4) *f has the standard shadowing property.*

Then Λ is hyperbolic.

Proof Let Λ be a compact f-invariant set, let $0 < \lambda < 1$ and $L > 0$ be given, and assume that f has properties (P.1)–(P.4). Let $T_\Lambda M = E \oplus F$ be a Df-invariant splitting as in (P.1) (recall that every dominated splitting is continuous). Thus, shrinking the neighborhood U of Λ, we may assume further that there exists an extension $T_{\Lambda_f(U)}M = \hat{E} \oplus \hat{F}$ of the dominated splitting $T_\Lambda M = E \oplus F$ (see Lemma 3.2.2).

Let us prove that Λ is hyperbolic. Assuming that E is not contracting, we show first that for any $\lambda < \eta < \eta' < 1$ there is $z \in \Lambda_f(U)$ such that

$$\liminf_{n\to\infty} \frac{1}{n} \sum_{j=0}^{n-1} \log \left\| Df_{|\hat{E}(f^j(z))} \right\| < \log \eta < \limsup_{n\to\infty} \frac{1}{n} \sum_{j=0}^{n-1} \log \left\| Df_{|\hat{E}(f^j(z))} \right\| < \log \eta'.$$

After that, we derive a contradiction by applying Proposition 3.2.3.

It is known that if there exists $N > 0$ such that for any $x \in \Lambda$ there is $0 \le n(x) \le N$ such that $\left\| Df^{n(x)}_{|E(x)} \right\| < 1$, then E is contracting.

Since E is not contracting, it is easy to see that there is $y_0 \in \Lambda$ such that

$$\prod_{j=0}^{n-1} \left\| Df_{|E(f^j(y_0))} \right\| \ge 1 \quad \text{for all} \quad n \ge 1$$

(recall that Λ is compact).

Choose $\varepsilon > 0$ small enough with $N(2\varepsilon, \Lambda) \subset U$ such that

(i) if $\text{dist}(x, y) < \varepsilon$ for some $x, y \in N(\varepsilon, \Lambda)$, then

$$\left| \log \|Df_{|\hat{E}(x)}\| - \log \|Df_{|\hat{E}(y)}\| \right| < \min \left\{ \frac{1}{2}(\log \eta' - \log \eta), \frac{1}{3}(\log \eta - \log \lambda) \right\}.$$

Observe that item (i) follows from the continuity of E (recall that $\hat{E}_{|\Lambda} = E$).

Since f has the standard shadowing property, there is $0 < \delta \le \varepsilon$ such that any δ-pseudotrajectory of f in M can be ε-shadowed by a trajectory of f.

Denote the ω-limit set of y_0 by $\omega_f(y_0)$. It is well known that $\omega_f(y_0) \subset \Lambda$ is an f-invariant compact set, and for any neighborhood $V = V(\omega_f(y_0))$ of $\omega_f(y_0)$ there is $N > 0$ such that $f^n(y_0) \in V$ for any $n \ge N$. By the compactness, there exists a finite set of points $\{x_j\}_{j=1}^{\ell}$ in $\omega_f(y_0)$ such that

$$\omega_f(y_0) \subset \bigcup_{j=1}^{\ell} N(\delta/2, x_j).$$

Since $P_{\dim E}(f)$ is dense in Λ, it is easy to see that for the chosen δ there exists a finite set of periodic points $\{p_j\}_{j=1}^{\ell} \subset P_{\dim E}(f)$ with $\text{dist}(x_j, p_j) < \frac{\delta}{2}$ such that

$$\omega_f(y_0) \subset \bigcup_{j=1}^{\ell} N(\delta, p_j)$$

and thus, there is $N' > 0$ such that

$$f^n(y_0) \in \bigcup_{j=1}^{\ell} N(\delta, p_j) \subset N(\varepsilon, \Lambda)$$

for any $n \ge N'$.

Assume that $n \ge N'$. Then

$$\prod_{j=0}^{n-1} \left\| Df_{|E(f^j(y_0))} \right\| = \prod_{j=N'}^{n-N'-1} \left\| Df_{|E(f^j(y_0))} \right\| \cdot \prod_{j=0}^{N'-1} \left\| Df_{|E(f^j(y_0))} \right\| \ge 1.$$

Thus,

$$\prod_{j=0}^{n-1} \left\| Df_{|E(f^{N'+j}(y_0))} \right\| \geq \left(\prod_{j=0}^{N'-1} \left\| Df_{|E(f^j(y_0))} \right\| \right)^{-1} \geq e^{-KN'},$$

so that

$$\frac{1}{n} \sum_{j=0}^{n-1} \log \left\| Df_{|E(f^{N'+j}(y_0))} \right\| \geq -\frac{KN'}{n}.$$

Here $K = \max \{ | \log \|Df(x)\| |, | \log \|Df^{-1}(x)\| | : x \in M \}$.

Hence,

$$(ii)\ \liminf_{n\to\infty} \frac{1}{n} \sum_{j=0}^{n-1} \log \left\| Df_{|E(f^{N'+j}(y_0))} \right\| \geq \lim_{n\to\infty} \left(-\frac{KN'}{n} \right) = 0.$$

We may assume that the period of p_j satisfies the inequality $\pi(p_j) \geq L$ for any j, and, finally, put

$$\pi = \prod_{j=1}^{\ell} \pi(p_j).$$

The set of periodic orbits

$$\mathscr{PO} = \bigcup_{j=1}^{\ell} O(p_j, f)$$

forms a δ-net of $\omega_f(y_0)$, i.e., for any $w \in \omega_f(y_0)$, there is $q \in \mathscr{PO}$ such that $\mathrm{dist}(w, q) < \delta$, and, conversely, for any $q \in \mathscr{PO}$, there is $w \in \omega_f(y_0)$ such that $\mathrm{dist}(w, q) < \delta$.

Observe that for any for any $q \in \mathscr{PO}$,

$$(iii)\ \frac{1}{\pi} \sum_{j=0}^{\pi-1} \log \left\| Df_{|E(f^j(q))} \right\| < \frac{1}{2}(\log \lambda + \log \eta)$$

by the choice of δ (see (P.2)).

We construct a δ-pseudotrajectory $\{x_i\}_{i\in\mathbb{Z}} \subset \Lambda$ of f composed of points of the orbit $O(y_0, f)$ and of the set $\mathscr{PO}$ by mimicking the procedure displayed in [109] (the construction is by induction). Denote $f^{N'}(y_0)$ by y_0 for simplicity.

Step I Since $y_0 \in \Lambda$, there is $q_{j_1} \in \mathscr{PO}$ such that $\mathrm{dist}(y_0, q_{j_1}) < \delta$. Set

$$x_{-1} = q_{j_1-1},\ x_{-2} = q_{j_1-2},\ \ldots,\ x_{-\pi+1} = q_{j_1-\pi+1},$$

$$x_{-\pi} = q_{j_1},\ x_{-\pi-1} = q_{j_1-1},\ x_{-\pi-2} = q_{j_1-2}\ \ldots.$$

Then $\mathrm{dist}(f(x_{-i}), x_{-i+1}) < \delta$ for $i \geq 1$, so that the negative part $\{x_i\}_{i=-\infty}^{-1}$ of $\{x_i\}_{i\in\mathbb{Z}}$ is constructed.

Step II Let $n_1 = 1$. Then

$$\frac{1}{n_1\pi}\left(n_1\sum_{j=0}^{\pi-1}\log\left\|Df_{|E(q_{j_1}+j)}\right\|\right) < \frac{1}{2}(\log\lambda + \log\eta).$$

Obviously, this inequality is ensured by *(iii)*.

Let $i_1 = n_1\pi$, put $x_j = q_{j_1+j}$ for $j = 0, 1, \ldots, i_1 - 1 = \pi - 1$, and put $x_{i_1} = y_0$. Then $\mathrm{dist}(f(x_j), x_{j+1}) < \delta$ for $j = 0, 1, \ldots, i_1 - 1$, and

$$\frac{1}{i_1}\sum_{j=0}^{i_1-1}\log\left\|Df_{|E(x_j)}\right\| < \frac{1}{2}(\log\lambda + \log\eta).$$

Put

$$a_j = \log\left\|Df_{|E(x_j)}\right\|$$

for $j = 0, 1, \ldots, i_1 - 1$, and choose l_1 so that

$$\frac{1}{i_1 + l_1}\left(\sum_{j=0}^{i_1-1}a_j + \sum_{j=0}^{l_1-1}\log\left\|Df_{|E(f^j(y_0))}\right\|\right) \geq \frac{1}{2}(\log\eta + \log\eta')$$

and

$$\frac{1}{i_1 + l}\left(\sum_{j=0}^{i_1-1}a_j + \sum_{j=0}^{l-1}\log\left\|Df_{|E(f^j(y_0))}\right\|\right) < \frac{1}{2}(\log\eta + \log\eta')$$

for any $l < l_1$.

The existence of l_1 is ensured by the choice of y_0 (recall the choice of y_0 and *(ii)*).

Set $j_1 = i_1 + l_1$, let $x_{i_1+1} = f(y_0), x_{i_1+2} = f^2(y_0), \ldots, x_{j_1-1} = f^{l_1-1}(y_0) \in O(y_0, f)$, and put

$$a_{i_1+j} = \log\left\|Df_{|E(x_{i_1}+j)}\right\|$$

for $j = 0, 1, \ldots, l_1 - 1$.

Step III Let i_{k-1}, j_{k-1}, $\{x_i\}_{i=0}^{j_{k-1}-1}$, and $\{a_i\}_{i=0}^{j_{k-1}-1}$ have been constructed in the previous steps. Similarly with the choice of q_{j_1} and n_1, we can choose $q_{j_k} \in \mathscr{P}\mathscr{O}$ so that

$$\operatorname{dist}\left(f(x_{j_{k-1}}), q_{j_k}\right) < \delta,$$

and a positive number n_k such that

$$\frac{1}{i_k}\left(\sum_{j=0}^{j_{k-1}-1} a_j + n_k \sum_{j=0}^{\pi-1} \log \left\| Df_{|E(q_{j_k}+j)} \right\| \right) < \frac{1}{2}(\log \lambda + \log \eta),$$

where $i_k = j_{k-1} + n_k \pi$ (the existence of n_k is ensured by (*iii*)). Let

$$x_{j_{k-1}+1} = q_{j_k+1},\ x_{j_{k-1}+2} = q_{j_k+2},\ \ldots,\ x_{j_{k-1}+\pi} = q_{j_k},$$

$$x_{j_{k-1}+\pi+1} = q_{j_k+1},\ x_{j_{k-1}+\pi+2} = q_{j_k+2},\ \ldots,$$

and $x_{i_k} = f(x_{j_{k-1}-1}) \in O(y_0, f)$.

Obviously,

$$\operatorname{dist}(f(x_{j_{k-1}+j}), x_{j_{k-1}+j+1}) < \delta$$

for $j = 0, 1, \ldots, n_k\pi - 1$. Put

$$a_{j_{k-1}+j} = \log \left\| Df_{|E(x_{j_{k-1}+j})} \right\|$$

for $j = 0, 1, \ldots, n_k \cdot \pi - 1$, and choose l_k so that

$$\frac{1}{i_k + l_k}\left(\sum_{j=0}^{i_k-1} a_j + \sum_{j=0}^{l_k-1} \log \left\| Df_{|E(f^j(x_{i_k}))} \right\| \right) \geq \frac{1}{2}(\log \eta + \log \eta'),$$

and

$$\frac{1}{i_k + l}\left(\sum_{j=0}^{i_k-1} a_j + \sum_{j=0}^{l} \log \left\| Df_{|E(f^j(x_{i_k}))} \right\| \right) < \frac{1}{2}(\log \eta + \log \eta')$$

for any $l < l_k$.

The existence of l_k is ensured by the fact that $x_{i_k} \in O(y_0, f)$ (recall the choice of y_0 and (*ii*)).

Let $j_k = i_k + l_k$ and let $x_{i_k+1} = f(x_{i_k}), x_{i_k+2} = f^2(x_{i_k}), \ldots, x_{j_k-1} = f^{l_k-1}(x_{i_k})$. Finally, we put

$$a_{j_{k-1}+j} = \log \left\| Df_{|E(f^j(x_{i_k}))} \right\|$$

for $j = 0, 1, \ldots, l_k - 1$.

This completes the construction of $\{x_i\}_{i\in\mathbb{Z}} \subset \Lambda$.
Roughly speaking, the δ-pseudotrajectory $\{x_i\}_{i\in\mathbb{Z}}$ looks as follows:

$$\{\cdots,\ \mathcal{PO},\ \mathcal{PO},\ y_0, f(y_0), f^2(y_0), \ldots, f^{l_1}(y_0),\ \mathcal{PO},$$

$$\ldots,\ \mathcal{PO}, f^{l_1+1}(y_0), \ldots, f^{l_1+l_2}(y_0),\ \mathcal{PO},\ \mathcal{PO}, \ldots\}.$$

Recall that $K = \max\left\{|\log\|Df(x)\||, |\log\|Df^{-1}(x)\|| : x \in M\right\}$.
It is easy to see that

$$\frac{1}{i_k}\sum_{j=0}^{i_{k-1}-1} a_j < \frac{1}{2}(\log\lambda + \log\eta) \quad\text{and}\quad \frac{1}{j_k}\sum_{j=0}^{j_k-1} a_j \geq \frac{1}{2}(\log\eta + \log\eta')$$

for every $k = 1, 2, \ldots,$ and

$$\frac{1}{n}\sum_{j=0}^{n-1} a_j < \frac{1}{n}\left(\frac{1}{2}(\log\eta + \log\eta')\,(n-\pi) + K\cdot\pi\right)$$

for every $n \geq \pi$.
Hence,

$$\limsup_{n\to\infty}\frac{1}{n}\sum_{j=0}^{n-1} a_j = \frac{1}{2}(\log\eta + \log\eta') \quad\text{and}\quad \liminf_{n\to\infty}\frac{1}{n}\sum_{j=0}^{n-1} a_j \leq \frac{1}{2}(\log\lambda + \log\eta).$$

Let $z \in M$ be a point whose f-trajectory ε-shadows $\{x_i\}_{i\in\mathbb{Z}}$ (see (P.4)). Note that $O(z,f) \subset N(2\varepsilon, \Lambda) \subset U$. Thus, by the choice of ε (see (i)),

$$\liminf_{n\to\infty}\frac{1}{n}\sum_{j=0}^{n-1}\log\left\|Df_{|\hat{E}(f^j(z))}\right\| < \log\eta < \limsup_{n\to\infty}\frac{1}{n}\sum_{j=0}^{n-1}\log\left\|Df_{|\hat{E}(f^j(z))}\right\| < \log\eta'.$$

By Proposition 3.2.3, there is a hyperbolic periodic point q of index $\dim E$ such that $O(q,f) \subset U$ and the derivatives along the trajectory $O(q,f)$ satisfy the inequalities

$$\prod_{i=0}^{k-1}\|Df_{|E^s(f^i(q))}\| \leq \eta'^k \quad\text{and}\quad \prod_{i=k-1}^{\pi(q)-1}\|Df_{|E^s(f^i(q))}\| > \eta^{\pi(q)-k+1}$$

for all $k = 1, 2, \ldots, \pi(q)$.

Furthermore, q can be chosen so that $\pi(q)$ is arbitrarily large, and thus we may assume that $\pi(q) \geq L$. This is a contradiction because

$$\prod_{i=0}^{\pi(q)-1} \|Df_{|E^s(f^i(q))}\| < \lambda^{\pi(q)}$$

by (P.2). Applying a similar reasoning, we can show that F is expanding, and thus, Λ is hyperbolic. □

Now we give a proof of a generalization of Liao's closing lemma (Lemma 3.2.4) proved by Gan [20].

Recall that a definition of a λ-quasi-hyperbolic orbit string

$$\{x, f(x), f^2(x), \cdots, f^n(x)\}$$

with respect to a splitting of $T_xM = E(x) \oplus F(x)$ has been given before Lemma 3.2.3.

Let $\{x_i\}_{i=-\infty}^{\infty}$ be a sequence of points in M and let $\{n_i\}_{i=-\infty}^{\infty}$ be a sequence of positive integers. Denote

$$\{x_i, n_i\} = \{f^j(x_i) : 0 \leq j \leq n_i - 1\}.$$

The sequence $\{x_i, n_i\}_{i=-\infty}^{\infty}$ is called a *λ-quasi-hyperbolic δ-pseudotrajectory* with respect to splittings $T_{x_i}M = E(x_i) \oplus F(x_i)$ if for any i, $\{x_i, n_i\}$ is λ-quasi-hyperbolic with respect to $T_{x_i}M = E(x_i) \oplus F(x_i)$ and $\text{dist}\,(f^{n_i}(x_i), x_{i+1}) \leq \delta$.

A point x *ε-shadows* $\{x_i, n_i\}_{i=-\infty}^{\infty}$ if

$$\text{dist}\left(f^j(x), f^{j-N_i}(x_i)\right) \leq \varepsilon \quad \text{for} \quad N_i \leq j \leq N_{i+1} - 1,$$

where N_i is defined as follows:

$$N_i = \begin{cases} 0, & \text{if } i = 0; \\ n_0 + n_1 + \cdots + n_{i-1}, & \text{if } i > 0; \\ n_i + n_{i+1} + \cdots + n_{-1} & \text{if } i < 0. \end{cases}$$

In the following result, it is assumed that Λ is a compact invariant set of $f \in \text{Diff}^1(M)$ and there is a continuous Df-invariant splitting $T_\Lambda M = E \oplus F$, i.e., $Df(x)(E(x)) = E(f(x))$ and $Df(x)(F(x)) = F(f(x))$.

Lemma 3.2.5 (Generalized Liao's Closing Lemma [20]) *For any $0 < \lambda < 1$ there exist $L > 0$ and $\delta_0 > 0$ such that for any $0 < \delta < \delta_0$ and any λ-quasi-hyperbolic δ-pseudotrajectory $\{x_i, n_i\}_{i=-\infty}^{\infty}$ with respect to the splitting $T_\Lambda M = E \oplus F$ there exists a point x that $L\delta$-shadows $\{x_i, n_i\}_{i=-\infty}^{\infty}$. Moreover, if the sequence $\{x_i, n_i\}_{i=-\infty}^{\infty}$ is periodic, i.e., there exists an $m > 0$ such that $x_{i+m} = x_i$ and $n_{i+m} = n_i$ for all i, then the point x can be chosen to be periodic with period N_m.*

Proof Let $(X, \|\cdot\|)$ be a Banach space and let

$$X(\eta) = \{v \in X : \|v\| \le \eta\}, \quad \eta > 0.$$

If X is the direct sum of two closed subspaces E and F, i.e., $X = E \oplus F$, then the angle between E and F is defined as

$$\angle(E, F) = \inf\{\|u - v\| : (u \in E, v \in F, \|u\| = 1) \text{ or } (u \in E, v \in F, \|v\| = 1)\}.$$

Since E and F are closed, $0 < \angle(E, F) \le 1$. □

The following lemma is well known (e.g., see [64]); we give a proof for completeness.

Lemma 3.2.6 *In the above notation, assume that* $X = E \oplus F$ *and* $\angle(E, F) \ge \alpha > 0$. *Let* $L : X \to X$ *be a linear automorphism of the form*

$$L = \begin{pmatrix} A & B \\ C & D \end{pmatrix} : E \oplus F \to E \oplus F$$

such that

$$\max\{\|A\|, \|D^{-1}\|\} \le \lambda \quad \text{and} \quad \max\{\|B\|, \|C\|\} \le \varepsilon$$

for some $0 < \lambda < 1$ *and* $\varepsilon > 0$.

If

$$\varepsilon_1 = \frac{2\varepsilon(1 + \lambda)}{\alpha^2(1 - \lambda)} < 1,$$

then $I - L$ *is invertible, and*

$$\left\|(I - L)^{-1}\right\| \le R = R(\lambda, \varepsilon, \alpha) = \frac{1 + \lambda}{\alpha(1 - \lambda)(1 - \varepsilon_1)}.$$

Proof Put $J = \begin{pmatrix} A & 0 \\ 0 & D \end{pmatrix}$ and $K = \begin{pmatrix} 0 & B \\ C & 0 \end{pmatrix}$. Then

$$(I - J)^{-1} = \begin{pmatrix} (I - A)^{-1} & 0 \\ 0 & (I - D)^{-1} \end{pmatrix},$$

$$\left\|(I - A)^{-1}\right\| \le \frac{1}{1 - \lambda}, \text{ and } \left\|(I - D)^{-1}\right\| \le \frac{\lambda}{1 - \lambda}.$$

If $u \in E$, $v \in F$ and $\|u + v\| = 1$, then, by the definition of $\angle(E, F)$,

$$1 = \|u + v\| \geq \angle(E, F)\|u\| \geq \alpha\|u\| \text{ and } \|u + v\| \geq \alpha\|v\|.$$

Thus,

$$\|(I - J)^{-1}(u + v)\| \leq \|(I - A)^{-1}u\| + \|(I - D)^{-1}v\| \leq \frac{1 + \lambda}{\alpha(1 - \lambda)},$$

and hence,

$$\|(I - J)^{-1}\| \leq \frac{1 + \lambda}{\alpha(1 - \lambda)}.$$

A similar reasoning shows that

$$\|K\| \leq \frac{2\varepsilon}{\alpha}.$$

Since

$$\varepsilon_1 = \frac{2\varepsilon(1 + \lambda)}{\alpha^2(1 - \lambda)} < 1$$

by assumption, $I - L = (I - J) - K = (I - J)(I - (I - J)^{-1}K)$ and $\|(I - J)^{-1}K\| \leq \varepsilon_1$. Hence, $I - L$ is invertible, and

$$\|(I - L)^{-1}\| = \|(I - J)^{-1}(I - (I - J)^{-1}K)^{-1}\| \leq R,$$

which proves our lemma. □

The sequence version of the shadowing lemma is derived from the following fixed point result. For completeness, we give a proof following the method of [64].

In the next proposition, we denote

$$R = R(\lambda, \epsilon, \alpha) = \frac{1 + \lambda}{\alpha(1 - \lambda)(1 - \epsilon_1)}, \quad L = 2R, \quad \text{and} \quad \delta_0 = \frac{\eta}{L}$$

for $0 < \lambda < 1$, $0 < \alpha \leq 1$, and $\epsilon > 0$ such that $\epsilon_1 = \frac{2\epsilon(1+\lambda)}{\alpha^2(1-\lambda)} < 1$ and $\eta > 0$.

The minimal Lipschitz constant of a map ϕ is denoted by $\mathrm{Lip}\,\phi$.

Proposition 3.2.5 *If $0 < \delta \leq \delta_0$ and $\Phi = N + \phi : X(\eta) \to X$, where N is a linear automorphism of the form*

$$N = \begin{pmatrix} A & B \\ C & D \end{pmatrix} : E \oplus F \to E \oplus F$$

such that

$$\max\{\|A\|, \|D^{-1}\|\} \leq \lambda,$$

$$\max\{\|B\|, \|C\|\} \leq \epsilon,$$

$\angle(E, F) \geq \alpha$, $\mathrm{Lip}\,\phi \leq \frac{1}{L}$, *and* $\|\phi(0)\| \leq \delta$, *then* Φ *has a unique fixed point* z *in* $X(\eta)$ *such that* $\|z\| \leq L\delta$.

Proof By Lemma 3.2.6, $I - N$ is invertible. Let

$$H = (I - N)^{-1}\phi : X(\eta) \to X.$$

The set of fixed points of H in $X(\eta)$ coincides with the set of fixed points of Φ in $X(\eta)$. If $x \in X(L\delta)$, then

$$\begin{aligned} \|H(x)\| &= \|H(0) + (H(x) - H(0))\| \leq \\ &\leq \|(I - N)^{-1}\phi(0)\| + \|(I - N)^{-1}(\phi(x) - \phi(0))\| \leq \\ &\leq R\delta + R\tfrac{1}{L}L\delta = L\delta. \end{aligned}$$

Thus, H maps $X(L\delta)$ to $X(L\delta)$.

If $x, y \in X(\eta)$, then

$$\|H(x) - H(y)\| = \|(I - N)^{-1}(\phi(x) - \phi(y))\| \leq R\frac{1}{L}\|x - y\| = \frac{\|x - y\|}{2}. \tag{3.7}$$

Hence, the map $H : X(L\delta) \to X(L\delta)$ is contracting. Therefore, H has a unique fixed point z in $X(L\delta)$. Moreover, if $z' \in X(\eta)$ is another fixed point of H, then $z = z'$ by (3.7). □

In the following proposition, let $X_i = \mathbb{R}^\nu$ for integer i (where $\nu = \dim M$) and we assume that $X_i = E_i \oplus F_i$. Let

$$Y = \prod_{i=-\infty}^{\infty} X_i$$

be endowed with the supremum norm

$$\|v\| = \sup\{|v_i|\}, \quad v = (v_i).$$

Thus, Y is a Banach space.

We consider a map $\Phi : Y \to Y$ of the form $(\Phi(v))_{i+1} = \Phi_i(v_i)$, where $\Phi_i : X_i \to X_{i+1}$.

Applying Proposition 3.2.5 to $\Phi : Y \to Y$, we obtain the sequence version of the shadowing lemma for hyperbolic pseudotrajectories in the following way.

Proposition 3.2.6 *Let us assume that conditions of Proposition 3.2.5 are satisfied and use the above notation.*

If $0 < \delta \le \delta_0$ and $\Phi : Y(\eta) \to Y$ has the form

$$\Phi_i = L_i + \phi_i : X_i(\eta) \to X_{i+1},$$

where

$$L_i = \begin{pmatrix} A_i & B_i \\ C_i & D_i \end{pmatrix}$$

with respect to the splitting $X_i = E_i \oplus F_i$ such that $\angle(E_i, F_i) \ge \alpha$,

$$\max\{\|A_i\|, \|D_i^{-1}\|\} \le \lambda, \quad \max\{\|B_i\|, \|C_i\|\} \le \epsilon, \quad \mathrm{Lip}\,\phi \le \frac{1}{L},$$

and $\|\phi_i(0)\| \le \delta$, then Φ has a unique fixed point $v \in Y(\eta)$, and $\|v\| \le L\delta$.

We need one more technical lemma. Fix $0 < \lambda < 1$.

A pair of sequences $\{a_i, b_i\}_{i=1}^n$ of positive numbers is called *λ-hyperbolic* if $a_k \le \lambda$ and $b_k \ge \lambda^{-1}$ for $k = 1, 2, \dots, n$.

A pair of sequences $\{a_i, b_i\}_{i=1}^n$ of positive numbers is called *λ-quasi-hyperbolic* if the following three conditions are satisfied:

(1) $\prod_{j=1}^k a_j \le \lambda^k$;
(2) $\prod_{j=k}^n b_j \ge \lambda^{k-n-1}$;
(3) $b_k/a_k \ge \lambda^{-2}$
for $k = 1, 2, \dots, n$.

A sequence $\{c_i\}_{i=1}^n$ of positive numbers is called a *balance sequence* if

$$\prod_{j=1}^k c_j \le 1 \quad \text{for} \quad k = 1, 2, \dots, n-1 \quad \text{and} \quad \prod_{j=1}^n c_j = 1.$$

A balance sequence $\{c_i\}_{i=1}^n$ is called *adapted to a λ-quasi-hyperbolic sequence pair* $\{a_i, b_i\}_{i=1}^n$ if $\{a_i/c_i, b_i/c_i\}_{i=1}^n$ is still λ-quasi-hyperbolic. Moreover, if $\{a_i/c_i, b_i/c_i\}_{i=1}^n$ is λ-hyperbolic, then $\{c_i\}_{i=1}^n$ is called *well adapted.*

If a balance sequence $\{c_i\}_{i=1}^n$ is adapted to a λ-quasi-hyperbolic sequence pair $\{a_i, b_i\}_{i=1}^n$, then we say that $\{a_i/c_i, b_i/c_i\}_{i=1}^n$ *is derived from* $\{a_i, b_i\}_{i=1}^n$. If $\{\bar{a}_i, \bar{b}_i\}_{i=1}^n$ is derived from $\{a_i, b_i\}_{i=1}^n$ and $\{\bar{\bar{a}}_i, \bar{\bar{b}}_i\}_{i=1}^n$ is derived from $\{\bar{a}_i, \bar{b}_i\}_{i=1}^n$, then $\{\bar{\bar{a}}_i, \bar{\bar{b}}_i\}_{i=1}^n$ is derived from $\{a_i, b_i\}_{i=1}^n$ as well.

Lemma 3.2.7 *Let* $0 < \lambda < 1$. *Then any λ-quasi-hyperbolic pair of sequences $\{a_i, b_i\}_{i=1}^n$ has a well adapted sequence $\{c_i\}_{i=1}^n$.*

Proof First we show that $\{a_i, b_i\}_{i=1}^n$ has an adapted sequence $\{c_i\}_{i=1}^n$ such that $a_i/c_i \le \lambda$ for $1 \le i \le n$.

To get a contradiction, assume that

$$N = \max\{k: \text{ there exist } \{c_i\}_{i=1}^n \text{ such that } a_i/c_i \leq \lambda, 1 \leq i \leq k\} < n.$$

Obviously, $N \geq 1$. Assume that $\{c_i\}_{i=1}^n$ is such an adapted sequence. Let $\bar{a}_i = a_i/c_i$, $\bar{b}_i = b_i/c_i$, $i = 1, 2, \dots, n$. Then $\bar{a}_{N+1} > \lambda$.

Since $\prod_{i=1}^{N+1} \bar{a}_i \leq \lambda^{N+1}$, there exists $1 \leq m < N+1$ such that

$$\prod_{i=k}^{N+1} \bar{a}_i > \lambda^{N+2-k} \quad \text{for} \quad k = m+1, \dots, N+1 \quad \text{and} \quad \prod_{i=m}^{N+1} \bar{a}_i \leq \lambda^{N+2-m}.$$

Let $\bar{c}_i = \bar{a}_i/\lambda$ for $i = m+1, \dots, N+1$ and $\bar{c}_i = 1$ for $i < m$ and $i > N+1$.

Then $\{\bar{c}_i\}_{i=1}^n$ is a balance sequence. Let $\bar{\bar{a}}_i = \bar{a}_i/\bar{c}_i$ and $\bar{\bar{b}}_i = \bar{b}_i/\bar{c}_i$ for $1 \leq i \leq n$ and put $\bar{c}_m = \left(\prod_{i=m+1}^{N+1} \bar{c}_i\right)^{-1}$. Obviously, $\bar{\bar{a}}_i = \lambda$ for $m+1 \leq i \leq N+1$,

$$\bar{\bar{a}}_m = \bar{a}_m/\bar{c}_m = \bar{a}_m \left(\prod_{i=m+1}^{N+1} \bar{c}_i\right) =$$
$$= \bar{a}_m \left(\prod_{i=m+1}^{N+1} \bar{a}_i\right) \lambda^{-(N-m+1)} = \left(\prod_{i=m}^{N+1} \bar{a}_i\right) \lambda^{-(N-m+1)} \leq \lambda,$$

and $\bar{\bar{b}}_i = \bar{b}_i/\bar{c}_i = \lambda \bar{b}_i/\bar{a}_i \geq \lambda^{-1}$ for $m+1 \leq i \leq N+1$.

Thus, one can easily check that $\{\bar{\bar{a}}_i, \bar{\bar{b}}_i\}_{i=1}^n$ is a λ-quasi-hyperbolic pair which is derived from $\{a_i, b_i\}_{i=1}^n$. But $\bar{\bar{a}}_i \leq \lambda$ for $1 \leq i \leq N+1$, which contradicts the maximality of N.

Similarly, $\{a_i, b_i\}_{i=1}^n$ has an adapted sequence $\{c_i\}_{i=1}^n$ such that $b_i/c_i \geq \lambda^{-1}$ for $1 \leq i \leq n$. In what follows, we assume that $\{a_i, b_i\}_{i=1}^n$ itself has the property that $b_i \geq \lambda^{-1}$ for $1 \leq i \leq n$. We will repeat the proof of the above paragraph to show that a well adapted sequence exists.

Let

$$N = \max\{k: \text{ there exist } \{c_i\}_{i=1}^n \text{ such that}$$
$$a_i/c_i \leq \lambda, 1 \leq i \leq k, \text{ and } b_i/c_i \geq \lambda, 1 \leq i \leq n\} < n.$$

Now we can copy the proof of the first paragraph word by word and only have to show that $\bar{\bar{b}}_m \geq \lambda^{-1}$. Since $\bar{c}_m \leq 1$, this is obvious. □

Remark 3.2.2 If $\{c_i\}_{i=1}^n$ is a well adapted sequence of $\{a_i, b_i\}_{i=1}^n$, then $a_i/c_i \leq \lambda$ and $b_i/c_i \geq \lambda^{-1}$. Hence, $a_i < a_i/\lambda \leq c_i \leq b_i\lambda < b_i$ for $i = 1, 2, \dots, n$.

We prove Lemma 3.2.5 (the generalized Liao's closing lemma) by combining Proposition 3.2.6 and Lemma 3.2.7.

Let $G_k(x)$, $x \in M$, be the Grassmann manifold of k-dimensional subspaces of the tangent space $T_x(M)$. Denote by $G_k(M)$ the bundle $\{G_k(x): x \in M\}$ and consider a metric ρ on $G_k(M)$ (we do not indicate the dependence of ρ on k).

The following lemma is an easy corollary of well-known properties of the exponential map.

Lemma 3.2.8 *For any $\alpha, \varepsilon, \tau, \gamma > 0$ there exists $\eta > 0$ such that if $x, y \in M$, $T_xM = E(x) \oplus F(x)$, $T_yM = E(y) \oplus F(y)$,*

$$\min\{\angle(E(x), F(x)), \angle(E_y, F_y)\} \geq \alpha,$$

and

$$\max\{\rho(Df(x)E(x), E(y)), \rho(Df(x)F(x), F(y)\} \leq \eta,$$

then the map

$$\Phi = \exp_y^{-1} of \circ \exp_x : T_xM(\eta) \to T_yM$$

can be written as $\Phi = L + \phi$, where
$L = \begin{pmatrix} A & B \\ C & D \end{pmatrix}$ *with respect to the splittings $E(x) \oplus F(x)$ and $E(y) \oplus F(y)$,*

$$1 - \tau \leq \frac{\|A\|}{\|Df|_{E(x)}\|} \leq 1 + \tau,$$

$$1 - \tau \leq \frac{\|D^{-1}\|^{-1}}{m\left(Df|_{F(x)}\right)} \leq 1 + \tau,$$

$$\max\{\|B\|, \|C\|\} \leq \varepsilon, \text{ and } \operatorname{Lip}\phi \leq \gamma.$$

Proof of Lemma 3.2.5 Let $\{x_i, n_i\}_{-\infty}^{\infty}$ be a λ-quasi-hyperbolic pseudotrajectory with respect to the splitting $T_\Lambda M = E \oplus F$. Denote

$$K = \sup_{x \in M} \{\|Df(x)\|, \|Df^{-1}(x)\|\} \quad \text{and} \quad \alpha = \inf_{x \in \Lambda} \angle(E(x), F(x)) > 0.$$

We first show that there exists a point z that ε-shadows $\{x_i, n_i\}_{i=-\infty}^{\infty}$, i.e.,

$$\operatorname{dist}\left(f^j(z), f^{j-N_i}(x_i)\right) \leq \varepsilon \quad \text{for} \quad N_i \leq j \leq N_{i+1} - 1,$$

where

$$N_i = \begin{cases} 0 & \text{if } i = 0; \\ n_0 + n_1 + \cdots + n_{i-1} & \text{if } i > 0; \\ n_i + n_{i+1} + \cdots + n_{-1} & \text{if } i < 0. \end{cases}$$

Let $y_j = f^{j-N_i}(x_i)$ for $N_i \le j < N_{i+1}$ and denote $X_j = T_{y_j}M$, $E_j = E(y_j)$, and $F_j = F(y_j)$.

Put $\mu = \frac{1+\lambda}{2}$ and $r = \mu/\lambda$ and take $\varepsilon > 0$ such that

$$\varepsilon_1 = \frac{2\varepsilon(1+\mu)}{\alpha^2(1-\mu)} < 1.$$

Let $R = R(\mu, \varepsilon, \alpha)$, $L = 2R$, and $\varepsilon_2 = \varepsilon/K$.

Since the splitting $T_\Lambda M = E \oplus F$ is continuous, it follows from Lemma 3.2.7 that if $\eta > 0$ is small enough and $\{y_j\}_{j=N_i}^{N_{i+1}}$ is λ-quasi-hyperbolic η-pseudotrajectory, then the map

$$\Phi_j = \exp_{y_{j+1}}^{-1} \circ f \circ \exp_{y_j} : X_j(\eta) \to X_{j+1}$$

has the form $\Phi_j = L_j + \phi_j$, where

$$L_j = \begin{pmatrix} A_j & B_j \\ C_j & D_j \end{pmatrix} : E_j \oplus F_j \to E_{j+1} \oplus F_{j+1}$$

and $\operatorname{Lip} \phi_j \le \frac{1}{KL}$.

If $N_i \le j < N_{i+1} - 1$, then $\phi_j(0) = 0$, $B_j = C_j = 0$, $A_j = Df|_{E_j}$, and $D_j = Df|_{F_j}$. If $j = N_{i+1} - 1$, then

$$\max\{\|B_j\|, \|C_j\|\} \le \varepsilon_2, \quad \|A_j\| \le r\|Df|_{E_j}\|, \text{ and } \|D_j^{-1}\| \le rm(Df|_{F_j})^{-1}.$$

Let $\delta_0 = \eta/L$ and fix $0 < \delta \le \delta_0$. If $\{x_i, n_i\}_{-\infty}^{\infty}$ is a quasi-hyperbolic δ-pseudotrajectory, then $\|\phi_j(0)\| \le \delta$. Thus, $\{\|A_j\|, m(D_j)\}_{j=N_i}^{N_{i+1}-1}$ is a μ-quasi-hyperbolic pair of sequences. Hence, there is a well adapted sequence $\{h_j\}_{j=N_i}^{N_{i+1}-1}$, i.e.,

$$\prod_{j=N_i}^{k} h_j \le 1 \quad \text{for } k = N_i, \dots, N_{i+1} - 2 \quad \text{and} \quad \prod_{j=N_i}^{N_{i+1}-1} h_j = 1,$$

where $\frac{1}{K} \le h_j \le K$.

Let $g_j = \prod_{k=N_i}^{j} h_k$, $\tilde{L}_j = h_j^{-1} L_j$, $\tilde{\phi}_j(x) = g_j^{-1}\phi_j(g_{j-1}(x))$ (note that $g_{N_i-1} = 1$), and $\tilde{\Phi}_j = \tilde{L}_j + \tilde{\phi}_j$. Denote

$$\Psi_j = \Phi_j \circ \cdots \circ \Phi_{N_i} \quad \text{and} \quad \tilde{\Psi}_j = \tilde{\Phi}_j \circ \cdots \circ \tilde{\Phi}_{N_i}.$$

Then $\tilde{\Psi}_j = g_j^{-1}\Psi_j$.

Note that $g_{N_{i+1}-1} = 1$ and $\tilde{\Psi}_{N_{i+1}-1} = \Psi_{N_{i+1}-1}$. Thus,

$$\text{Lip}\,\tilde{\phi}_j = g_j^{-1}\text{Lip}\,\phi_j g_{j-1} = h_j^{-1}\text{Lip}\,\phi_j \le K\frac{1}{KL} = \frac{1}{L},$$

$\tilde{\phi}_j(0) = \phi_j(0) = 0$ for $j = N_i, \dots, N_{i+1} - 2$, and $\tilde{\phi}_j(0) = g_j^{-1}\phi_j(0) = \phi_j(0)$ for $j = N_{i+1} - 1$ since $g_j = 1$.

Hence, by Proposition 3.2.6, $\tilde{\Phi} = \{\tilde{\Phi}_j\} : Y(\eta) \to Y$ (where $Y = \prod_{i=-\infty}^{\infty} X_i$) has a unique fixed point $\tilde{v} = \{\tilde{v}_j\}$, and $\|\tilde{v}\| \le L\delta$. Let $v_{N_i} = \tilde{v}_{N_i}$ and for $N_i < j < N_{i+1} - 1$, define $v_j = \Phi_{j-1}(v_{j-1})$ inductively.

To guarantee that this is possible, let us check that $\|v_j\| \le L\delta$. Since

$$v_j = \Psi_{j-1}(v_{N_i}) = g_{j-1}\tilde{\Psi}_{j-1} = g_{j-1}\tilde{v}_j,$$

we have the inequalities $\|v_j\| \le \|\tilde{v}_j\| \le L\delta$.

Since

$$v_{N_{i+1}} = \tilde{v}_{N_{i+1}} = \tilde{\Psi}_{N_{i+1}-1}(v_{N_i}) = \Psi_{N_{i+1}-1}(v_{N_i}) = \Phi_{N_{i+1}-1}(v_{N_{i+1}-1}),$$

v is a fixed point of Φ, and $\|v\| \le L\delta$. Then the f-trajectory of the point $z = \exp_{y_0}(v_0)$ $L\delta$-shadows $\{y_j\}$. This proves the first conclusion of Lemma 3.2.5.

Now we assume that the sequence $\{x_i, n_i\}_{i=-\infty}^{\infty}$ is periodic, i.e., there exists an $m > 0$ such that $x_{i+m} = x_i$ and $n_{i+m} = n_i$ for all i.

Define $\tilde{w}$ by $(\tilde{w})_i = (\tilde{v})_{N_m+i}$. Since $\tilde{v}$ and $\tilde{w}$ are fixed points of $\tilde{\Phi}$ in $Y(L\delta)$, $\tilde{v} = \tilde{w}$ by Proposition 3.2.6. Thus, $v = w$, and z has period N_m. □

Historical Remarks The theory involving a selection of some special kinds of λ-quasi-hyperbolic strings has its origins in the works of V. A. Pliss [73] and S. T. Liao [36].

The notion of λ-quasi-hyperbolic string and Liao's closing lemma played an essential part in the solution of the stability conjecture in [45].

3.3 Vector Fields in $\text{Int}^1(\text{OrientSP}_F \setminus \mathscr{B})$

To formulate our main results in the last two sections of Chap. 3, we need one more definition.

Consider a smooth vector field X on a smooth closed manifold M.

Let us say that a vector field X belongs to the *class* $\mathscr{B}$ if X has two hyperbolic rest points p and q (not necessarily different) with the following properties:

(1) The Jacobi matrix $DX(q)$ has two complex conjugate eigenvalues $\mu_{1,2} = a_1 \pm ib_1$ of multiplicity one with $a_1 < 0$ such that if $\lambda \ne \mu_{1,2}$ is an eigenvalue of $DX(q)$ with $\text{Re}\lambda < 0$, then $\text{Re}\lambda < a_1$;

(2) the Jacobi matrix $DX(p)$ has two complex conjugate eigenvalues $\nu_{1,2} = a_2 \pm ib_2$ with $a_2 > 0$ of multiplicity one such that if $\lambda \neq \nu_{1,2}$ is an eigenvalue of $DX(p)$ with $\text{Re}\lambda > 0$, then $\text{Re}\lambda > a_2$;
(3) the stable manifold $W^s(p)$ and the unstable manifold $W^u(q)$ have a trajectory of nontransverse intersection.

Clearly, vector fields $X \in \mathscr{B}$ are not structurally stable.

Condition (1) above means that the "weakest" contraction in $W^s(q)$ is due to the eigenvalues $\mu_{1,2}$ (condition (2) has a similar meaning).

The main result of this section is as follows.

Theorem 3.3.1

$$Int^1(OrientSP_F \setminus \mathscr{B}) \subset \mathscr{S}_F. \tag{3.8}$$

It follows from Theorem 1.4.1 (2) that $\mathscr{S}_F \subset \text{SSP}_F$; since the set $\mathscr{S}_F$ is C^1-open and $\mathscr{S}_F \cap \mathscr{B} = \emptyset$,

$$\mathscr{S}_F \subset \text{Int}^1(\text{SSP}_F \setminus \mathscr{B}) \subset \text{Int}^1(\text{OrientSP}_F \setminus \mathscr{B}).$$

Combining this inclusion with (3.8), we see that

$$\text{Int}^1(\text{OrientSP}_F \setminus \mathscr{B}) = \mathscr{S}_F.$$

Proof The proof of inclusion (3.8) is based on Theorem 1.3.13 (2):

$$\text{Int}^1(\text{KS}_F) = \mathscr{S}_F$$

(recall that KS_F is the set of Kupka–Smale vector fields).

Thus, in fact, we are going to prove that

$$\text{Int}^1(\text{OrientSP}_F \setminus \mathscr{B}) \subset \text{KS}_F. \tag{3.9}$$

Before proving inclusion (3.9), we introduce some terminology and notation.

The term "transverse section" will mean a smooth open disk in M of codimension 1 that is transverse to the flow ϕ at any of its points.

Let, as above, $\text{Per}(X)$ denote the set of rest points and closed orbits of a vector field X.

Recall (see Sect. 1.3) that we have denoted by HP_F the set of vector fields X for which any trajectory of the set $\text{Per}(X)$ is hyperbolic. Our first lemma is valid for the set OrbitSP_F (which is, in general, larger than OrientSP_F); we prove it in this, more general form, since it can be applied for other purposes.

Lemma 3.3.1

$$Int^1(OrbitSP_F) \subset HP_F. \tag{3.10}$$

Proof To get a contradiction, let us assume that there exists a vector field $X \in \mathrm{Int}^1(\mathrm{OrbitSP}_F)$ that does not belong to HP_F, i.e., the set $\mathrm{Per}(X)$ contains a trajectory p that is not hyperbolic.

Let us first consider the case where p is a rest point. Identify M with $\mathbb{R}^n$ in a neighborhood of p. Applying an arbitrarily C^1-small perturbation of the field X, we can find a field $Y \in \mathrm{Int}^1(\mathrm{OrbitSP}_F)$ that is linear in a neighborhood U of p (we also assume that p is the origin of U).

(Here and below in the proof of Lemma 3.3.1, all the perturbations are C^1-small perturbations that leave the field in $\mathrm{Int}^1(\mathrm{OrbitSP}_F)$; we denote the perturbed fields by the same symbol X and their flows by ϕ.)

Then trajectories of X in U are governed by a differential equation

$$\dot{x} = Px, \tag{3.11}$$

where the matrix P has an eigenvalue λ with $\mathrm{Re}\lambda = 0$.

Consider first the case where $\lambda = 0$. We perturb the field X (and change coordinates, if necessary) so that, in Eq. (3.11), the matrix P is block-diagonal,

$$P = \mathrm{diag}(0, P_1), \tag{3.12}$$

and P_1 is an $(n-1) \times (n-1)$ matrix.

Represent coordinate x in U as $x = (y, z)$ with respect to (3.12); then

$$\phi(t, (y, z)) = (y, \exp(P_1 t)z)$$

in U.

Take $\varepsilon > 0$ such that $N(4\varepsilon, p) \subset U$. To get a contradiction, assume that $X \in \mathrm{OrbitSP}$; let d correspond to the chosen ε.

Fix a natural number m and consider the following mapping from $\mathbb{R}$ into U:

$$g(t) = \begin{cases} y = -2\varepsilon, & z = 0; \quad t \le 0; \\ y = -2\varepsilon + t/m, & z = 0; \quad 0 < t < 4m\varepsilon; \\ y = 2\varepsilon, & z = 0; \quad 4m\varepsilon < t. \end{cases}$$

Since the mapping g is continuous, piecewise differentiable, and either $\dot{y} = 0$ or $\dot{y} = 1/m$, g is a d-pseudotrajectory for large m.

Any trajectory of X in U belongs to a plane $y = \text{const}$; hence,

$$\mathrm{dist}_H\big(\mathrm{Cl}(O(q, \phi)), \mathrm{Cl}(\{g(t) : t \in \mathbb{R}\})\big) \ge 2\varepsilon$$

for any q. This completes the proof in the case considered.

A similar reasoning works if p is a rest point and the matrix P in (3.12) has a pair of eigenvalues $\pm ib, b \neq 0$.

Now we assume that p is a nonhyperbolic closed trajectory. In this case, we perturb the vector field X in a neighborhood of the trajectory p using the perturbation technique developed by Pugh and Robinson in [77]. Let us formulate their result (which will be used below several times).

Pugh-Robinson Pertubation *Assume that r_1 is not a rest point of a vector field X. Let $r_2 = \phi(\tau, r_1)$, where $\tau > 0$. Let Σ_1 and Σ_2 be two small transverse sections such that $r_i \in \Sigma_i, i = 1, 2$. Let σ be the local Poincaré transformation generated by these transverse sections.*

Consider a point $r' = \phi(\tau', r_1)$, where $\tau' \in (0, \tau)$, and let U be an arbitrary open set containing r'.

Fix an arbitrary C^1 neighborhood F of the field X.

There exist positive numbers ε_0 and Δ_0 with the following property: if σ' is a local diffeomorphism from the Δ_0-neighborhood of r_1 in Σ_1 into Σ_2 such that

$$dist_{C^1}(\sigma, \sigma') < \varepsilon_0,$$

then there exists a vector field $X' \in F$ such that

(1) *$X' = X$ outside U;*
(2) *σ' is the local Poincaré transformation generated by the sections Σ_1 and Σ_2 and trajectories of the field X'.*

Let ω be the least positive period of the nonhyperbolic closed trajectory p. We fix a point $\pi \in p$, local coordinates in which π is the center, and a hyperplane Σ of codimension 1 transverse to the vector $F(\pi)$. Let y be coordinate in Σ.

Let σ be the local Poincaré transformation generated by the transverse section Σ; denote $P = D\sigma(0)$. Our assumption implies that the matrix P is not hyperbolic. In an arbitrarily small neighborhood of the matrix P, we can find a matrix P' such that P' either has a real eigenvalue with unit absolute value of multiplicity 1 or a pair of complex conjugate eigenvalues with unit absolute value of multiplicity 1. In both cases, we can choose coordinates $y = (v, w)$ in Σ in which

$$P' = \text{diag}(Q, P_1), \tag{3.13}$$

where Q is a 1×1 or 2×2 matrix such that $|Qv| = |v|$ for any v.

Now we can apply the Pugh-Robinson perturbation (taking $r_1 = r_2 = \pi$ and $\Sigma_1 = \Sigma_2 = \Sigma$) which modifies X in a small neighborhood of the point $\phi(\omega/2, \pi)$ and such that, for the perturbed vector field X', the local Poincaré transformation generated by the transverse section Σ is given by $y \mapsto P'y$.

Clearly, in this case, the trajectory of π in the field X' is still closed (with some period ω'). As was mentioned, we assume that X' has the orbital shadowing property (and write X, ϕ, ω instead of X', ϕ', ω').

We introduce in a neighborhood of the point π coordinates $x = (x', y)$, where x' is one-dimensional (with axis parallel to $X(\pi)$), and y has the above-mentioned property.

Of course, the new coordinates generate a new metric, but this new metric is equivalent to the original one; thus, the corresponding shadowing property (or its absence) is preserved.

We need below one more technical statement.

LE (Local Estimate) *There exists a neighborhood W of the origin in Σ and constants $l, \delta_0 > 0$ with the following property: If $z_1 \in \Sigma \cap W$ and $|z_2 - z_1| < \delta < \delta_0$, then we can represent z_2 as $\phi(\tau, z_2')$ with $z_2' \in \Sigma$ and*

$$|\tau|,\ |z_2' - z_1| < l\delta. \tag{3.14}$$

This statement is an immediate corollary of the theorem on local rectification of trajectories (see, for example, [8]): In a neighborhood of a point that is not a rest point, the flow of a vector field of class C^1 is diffeomorphic to the family of parallel lines along which points move with unit speed (and it is enough to note that a diffeomorphic image of Σ is a smooth submanifold transverse to lines of the family).

We may assume that the neighborhood W in LE is so small that for $y \in \Sigma \cap W$, the function $\alpha(y)$ (the time of first return to Σ) is defined, and that the point $\phi(\alpha(v, w), (0, v, w))$ has coordinates $(Qv, P_1 w)$ in Σ.

Let us take a neighborhood U of the trajectory p such that if $r \in U$, then the first point of intersection of the positive semitrajectory of r with Σ belongs to W.

Take $a > 0$ such that the $4a$-neighborhood of the origin in Σ is a subset of W. Fix

$$\varepsilon < \min\left(\delta_0, \frac{a}{4l}\right),$$

where δ_0 and l satisfy the LE. Let d correspond to this ε (in the definition of the orbital shadowing property).

Take $y_0 = (v_0, 0)$ with $|v_0| = a$. Fix a natural number ν and set

$$\alpha_k = \alpha\left(\left(\frac{k}{\nu} Q^k v_0, 0\right)\right), \quad k \in [0, \nu - 1),$$

$$\beta_0 = 0, \quad \beta_k = \alpha_1 + \cdots + \alpha_k,$$

and

$$g(t) = \begin{cases} \phi(t, (0, 0, 0)), & t < 0; \\ \phi\left(t - \beta_k, \left(0, \frac{k}{\nu} Q^k v_0, 0\right)\right), & \beta_k \le t < \beta_{k+1},\ k \in [0, \nu - 1); \\ \phi\left(t - \beta_\nu, (0, Q^\nu v_0, 0)\right), & t \ge \beta_\nu. \end{cases}$$

Note that for any point $y = (v, 0)$ of intersection of the set $\{g(t) : t \in \mathbb{R}\}$ with Σ, the inequality $|v| \le a$ holds. Hence, we can take a so small that

$$N(2a, \mathrm{Cl}(\{g(t) : t \in \mathbb{R}\})) \subset U.$$

Since

$$\left|\frac{k}{\nu}Q^{k+1}v_0 - \frac{k+1}{\nu}Q^{k+1}v_0\right| = \frac{a}{\nu} \to 0, \quad \nu \to \infty,$$

$g(t)$ is a d-pseudotrajectory for large ν.

Assume that there exists a point q such that

$$\text{dist}_H(\text{Cl}(O(q,\phi)), \text{Cl}(\{g(t) : t \in \mathbb{R}\})) < \epsilon.$$

In this case, $O(q,\phi) \subset U$, and there exist points $q_1, q_2 \in O(q,\phi)$ such that

$$|q_1| = |q_1 - (0,0,0)| < \varepsilon$$

and

$$|q_2 - (0, Q^{\nu}v_0, 0)| < \varepsilon.$$

By the choice of ε, there exist points $q_1', q_2' \in O(q,\phi) \cap \Sigma$ such that

$$|q_1'| < l\varepsilon < a/4 \quad \text{and} \quad |q_2' - Q^{\nu}v_0| < l\varepsilon < a/4.$$

Let $q_1' = (0, v_1, w_1)$ and $q_2' = (0, v_2, w_2)$. Since these points belong to the same trajectory that is contained in U, $|v_1| = |v_2|$. At the same time,

$$|v_1| < a/4, \quad |v_2 - Q^{\nu}v_0| < a/4, \quad \text{and} \quad |Q^{\nu}v_0| = a,$$

and we get a contradiction which proves Lemma 3.3.1. □

To complete the proof of Theorem 3.3.1, we show that any vector field

$$X \in \text{Int}^1(\text{OrientSP}_F \setminus \mathscr{B})$$

has the second property from the definition of Kupka–Smale flows, i.e., stable and unstable manifolds of trajectories of the set $\text{Per}(X)$ are transverse.

Then

$$\text{Int}^1(\text{OrientSP}_F \setminus \mathscr{B}) \subset \text{KS}_F;$$

hence, inclusion (3.9) is valid.

To get a contradiction, let us assume that there exist trajectories $p, q \in \text{Per}(X)$ for which the unstable manifold $W^u(q)$ and the stable manifold $W^s(p)$ have a point r of nontransverse intersection. We have to consider separately the following two cases.

Case (B1): p and q are rest points of the flow ϕ.
Case (B2): either p or q is a closed trajectory.

Case (B1) Since $X \notin \mathscr{B}$, we may assume (after an additional perturbation, if necessary) that the eigenvalues $\lambda_1, \ldots, \lambda_u$ with $\mathrm{Re}\lambda_j > 0$ of the Jacobi matrix $DX(p)$ have the following property:

$$\mathrm{Re}\lambda_j > \lambda_1 > 0, \quad j = 2, \ldots, u$$

(where u is the dimension of $W^u(p)$). This property means that there exists a one-dimensional "direction of weakest expansion" in $W^u(p)$.

If this is not the case, then our assumption that $X \notin \mathscr{B}$ implies that the eigenvalues $\mu_1, \ldots, \mu_s$ with $\mathrm{Re}\mu_j < 0$ of the Jacobi matrix $DX(q)$ have the following property:

$$\mathrm{Re}\mu_j < \mu_1 < 0, \quad j = 2, \ldots, s$$

(where s is the dimension of $W^s(q)$). If this condition holds, we reduce the problem to the previous case by passing from the field X to the field $-X$ (clearly, the fields X and $-X$ have the oriented shadowing property simultaneously).

Making a perturbation (in this part of the proof, we always assume that the perturbed field belongs to the set $\mathrm{OrientSP} \setminus \mathscr{B}$), we may "linearize" the field X in a neighborhood U of the point p; thus, trajectories of X in U are governed by a differential equation

$$\dot{x} = Px,$$

where

$$P = \mathrm{diag}(P_s, P_u), \quad P_u = \mathrm{diag}(\lambda, P_1), \quad \lambda > 0, \tag{3.15}$$

P_1 is a $(u-1) \times (u-1)$ matrix for which there exist constants $K > 0$ and $\mu > \lambda$ such that

$$\|\exp(-P_1 t)\| \le K^{-1} \exp(-\mu t), \quad t \ge 0, \tag{3.16}$$

and $\mathrm{Re}\lambda_j < 0$ for the eigenvalues λ_j of the matrix P_s.

Let us explain how to perform the above-mentioned perturbations preserving the nontransversality of $W^u(q)$ and $W^s(p)$ at the point r (we note that a similar reasoning can be used in "replacement" of a component of intersection of $W^u(q)$ with a transverse section Σ by an affine space, see the text preceding Lemma 3.3.2 below).

Consider points $r^* = \phi(\tau, r)$, where $\tau > 0$, and $r' = \phi(\tau', r)$, where $\tau' \in (0, \tau)$. Let Σ and Σ^* be small transverse sections that contain the points r and r^*. Take small neighborhoods V and U' of p and r', respectively, so that the set V does not intersect the "tube" formed by pieces of trajectories through points of U' whose endpoints belong to Σ and Σ^*. In this case, if we perturb the vector field X in V and apply the Pugh-Robinson perturbation in U', these perturbations are "independent."

We perturb the vector field X in V obtaining vector fields X' that are linear in small neighborhoods $V' \subset V$ and such that the values $\rho_1(X, X')$ are arbitrarily small.

Let γ_s and γ_s^* be the components of intersection of the stable manifold $W^s(p)$ (for the field X) with Σ and Σ^* that contain the points r and r^*, respectively.

Since the stable manifold of a hyperbolic rest point depends (on its compact subsets) C^1-smoothly on C^1-small perturbations, the stable manifolds $W^s(p)$ (for the perturbed fields X') contain components γ_s' of intersection with Σ^* that converge (in the C^1 metric) to γ_s^*.

Now we apply the Pugh-Robinson perturbation in U' and find a field X' in an arbitrary C^1 neighborhood of X such that the local Poincaré transformation generated by the field X' and sections Σ and Σ^* takes γ_s' to γ_s (which means that the nontransversality at r is preserved).

We introduce in U coordinates $x = (y; v, w)$ according to (3.15): y is coordinate in the s-dimensional "stable" subspace (denoted E^s); (v, w) are coordinates in the u-dimensional "unstable" subspace (denoted E^u). The one-dimensional coordinate v corresponds to the eigenvalue λ (and hence to the one-dimensional "direction of weakest expansion" in E^u).

In the neighborhood U,

$$\phi(t, (y, v, w)) = (\exp(P_s t)y; \exp(\lambda t)v, \exp(P_1 t)w),$$

and it follows from (3.16) that

$$|\exp(P_1 t)w| \geq K \exp(\mu t)|w|, \quad t \geq 0. \tag{3.17}$$

Denote by E_1^u the one-dimensional invariant subspace corresponding to λ.

We naturally identify $E^s \cap U$ and $E^u \cap U$ with the intersections of U with the corresponding local stable and unstable manifolds of p, respectively.

Let us construct a special transverse section for the flow ϕ. We may assume that the point r of nontransverse intersection of $W^u(q)$ and $W^s(p)$ belongs to U. Take a hyperplane Σ' in E^s of dimension $s - 1$ that is transverse to the vector $X(r)$. Set $\Sigma = \Sigma' + E^u$; clearly, Σ is transverse to $X(r)$.

By a perturbation of the field X outside U, we may get the following: in a neighborhood of r, the component of intersection $W^u(q) \cap \Sigma$ containing r (for the perturbed field) has the form of an affine space $r + L$, where L is the tangent space, $L = T_r(W^u(q) \cap \Sigma)$, of the intersection $W^u(q) \cap \Sigma$ at the point r for the unperturbed field (compare, for example, with [33]).

Let Σ_r be a small transverse disk in Σ containing the point r. Denote by γ the component of intersection of $W^u(q) \cap \Sigma_r$ containing r.

Lemma 3.3.2 *There exists $\varepsilon > 0$ such that if $x \in \Sigma_r$ and*

$$dist(\phi(t, x), O^-(r, \phi)) < \varepsilon, \quad t \leq 0, \tag{3.18}$$

then $x \in \gamma$.

Proof To simplify presentation, let us assume that q is a rest point; the case of a closed trajectory is considered using a similar reasoning.

By the Grobman–Hartman theorem, there exists $\varepsilon_0 > 0$ such that the flow of X in $N(2\varepsilon_0, q)$ is topologically conjugate to the flow of a linear vector field.

Denote by A the intersection of the local stable manifold of q, $W^s_{loc}(q)$, with the boundary of the ball $N(2\varepsilon_0, q)$.

Take a negative time T such that if $s = \phi(T, r)$, then

$$\phi(t, s) \in N(\varepsilon_0, q), \quad t \leq 0. \tag{3.19}$$

Clearly, if ε_0 is small enough, then the compact sets A and

$$B = \{\phi(t, r) : T \leq t \leq 0\}$$

are disjoint. There exists a positive number $\varepsilon_1 < \varepsilon_0$ such that the ε_1-neighborhoods of the sets A and B are disjoint as well.

Take $\varepsilon_2 \in (0, \varepsilon_1)$. There exists a neighborhood V of the point s with the following property: If $y \in V \setminus W^u_{loc}(q)$, then the first point of intersection of the negative semitrajectory of y with the boundary of $N(2\varepsilon_0, q)$ belongs to the ε_2-neighborhood of the set A (this statement is obvious for a neighborhood of a saddle rest point of a linear vector field; by the Grobman-Hartman theorem, it holds for X as well).

Clearly, there exists a small transverse disk Σ_s containing s and such that if $y \in \Sigma_s \cap W^u_{loc}(q)$, then the first point of intersection of the positive semitrajectory of y with the disk Σ_r belongs to γ (in addition, we assume that Σ_s belongs to the chosen neighborhood V).

There exists $\varepsilon \in (0, \varepsilon_1 - \varepsilon_2)$ such that the flow of X generates a local Poincaré transformation

$$\sigma : \Sigma_r \cap N(\varepsilon, r) \to \Sigma_s.$$

Let us show that this ε has the desired property. It follows from our choice of Σ_s and (3.18) with $t = 0$ that if $x \notin \gamma$, then

$$y := \sigma(x) \in \Sigma_s \setminus W^u_{loc}(q);$$

in this case, there exists $\tau < 0$ such that the point $z = \phi(\tau, y)$ belongs to the intersection of $N(\varepsilon_2, A)$ with the boundary of $N(2\varepsilon_0, q)$. By (3.19),

$$\mathrm{dist}(z, \phi(t, s)) > \varepsilon_0, \quad t \leq 0. \tag{3.20}$$

At the same time,

$$\mathrm{dist}(z, \phi(t, r)) > \varepsilon_1 - \varepsilon_2, \quad T \leq t \leq 0. \tag{3.21}$$

Inequalities (3.20) and (3.21) contradict condition (3.18). Our lemma is proved. □

Now let us formulate the property of nontransversality of $W^u(q)$ and $W^s(p)$ at the point r in terms of the introduced objects. Recall that we work in a small neighborhood U of the rest point p identified with the Euclidean space $\mathbb{R}^n$.

Let Π^u be the projection to E^u parallel to E^s.

The transversality of $W^u(q)$ and $W^s(p)$ at r means that

$$T_r W^u(q) + T_r W^s(p) = \mathbb{R}^n.$$

Since Σ is a transverse section to the flow ϕ at r, the above equality is equivalent to the equality

$$L + E^s = \mathbb{R}^n.$$

Thus, the nontransversality means that

$$L + E^s \neq \mathbb{R}^n,$$

which implies that

$$L' := \Pi^u L \neq E^u. \tag{3.22}$$

We claim that there exists a linear isomorphism J of Σ for which the norm $\|J - \mathrm{Id}\|$ is arbitrarily small and such that

$$\Pi^u J L \cap E_1^u = \{0\}. \tag{3.23}$$

Let e be a unit vector of the line E_1^u. If $e \notin L'$, we have nothing to prove (take $J = \mathrm{Id}$). Thus, we assume that $e \in L'$. Since $L' \neq E^u$, there exists a vector $v \in E^u \setminus L'$.

Fix a natural number N and consider a unit vector v_N that is parallel to $Ne + v$. Clearly, $v_N \to e$ as $N \to \infty$. There exists a sequence T_N of linear isomorphisms of E^u such that $T_N v_N = e$ and

$$\|T_N - \mathrm{Id}\| \to 0, \quad N \to \infty.$$

Note that $T_N^{-1} e$ is parallel to v_N; hence, $T_N^{-1} e$ does not belong to L', and

$$T_N \Pi^u L \cap E_1^u = \{0\}. \tag{3.24}$$

Define an isomorphism J_N of Σ by

$$J_N(y, z) = (y, T_N z)$$

and note that

$$\|J_N - \mathrm{Id}\| \to 0, \quad N \to \infty.$$

Let $L_N = J_N L$. Equality (3.24) implies that

$$\Pi^u L_N \cap E_1^u = \{0\}. \tag{3.25}$$

Our claim is proved.

First we consider the case where $\dim E^u \geq 2$. Since $\dim L' < \dim E^u$ by (3.22) and $\dim E_1^u = 1$, our reasoning above (combined with a Pugh-Robinson perturbation) shows that we may assume that

$$L' \cap E_1^u = \{0\}. \tag{3.26}$$

For this purpose, we take a small transverse section Σ' containing the point $r' = \phi(-1, r)$, denote by γ the component of intersection of $W^u(q)$ with Σ' containing r', and note that the local Poincaré transformation σ generated by Σ' and Σ takes γ to the linear space L (in local coordinates of Σ). The mapping $\sigma_N = J_N \sigma$ is C^1-close to σ for large N and takes γ to L_N for which equality (3.25) is valid. Thus, we get equality (3.26) for the perturbed vector field.

This equality implies that there exists a constant $C > 0$ such that if $(y; v, w) \in r + L$, then

$$|v| \leq C|w|. \tag{3.27}$$

Fix $a > 0$ such that $N(4a, p) \subset U$. Take a point $\alpha = (0; a, 0) \in E_1^u$ and a positive number T and set $\alpha_T = (r_y; a\exp(-\lambda T), 0)$, where r_y is the y-coordinate of r. Construct a pseudotrajectory as follows:

$$g(t) = \begin{cases} \phi(t, r), & t \leq 0; \\ \phi(t, \alpha_T), & t > 0. \end{cases}$$

Since

$$|r - \alpha_T| = a\exp(-\lambda T) \to 0$$

as $T \to \infty$, for any d there exists T such that g is a d-pseudotrajectory.

Lemma 3.3.3 *Assume that $b \in (0, a)$ satisfies the inequality*

$$\log K - \log C + \left(\frac{\mu}{\lambda} - 1\right)\left(\log\frac{a}{2} - \log b\right) \geq 0.$$

Then for any $T > 0$, reparametrization h, and a point $s \in r + L$ such that $|r - s| < b$ there exists $\tau \in [0, T]$ such that

$$|\phi(h(\tau), s) - g(\tau)| \geq \frac{a}{2}.$$

Proof To get a contradiction, assume that

$$|\phi(h(\tau), s) - g(\tau)| < \frac{a}{2}, \quad \tau \in [0, T]. \tag{3.28}$$

Let $s = (y_0; v_0, w_0) \in r + L$. Since $|r - s| < b$,

$$|v_0| < b. \tag{3.29}$$

By (3.28),

$$\phi(h(\tau), s) \in U, \quad \tau \in [0, T].$$

Take $\tau = T$ in (3.28) to show that

$$|v_0| \exp(\lambda h(T)) > \frac{a}{2}.$$

It follows that

$$h(T) > \lambda^{-1} \left(\log \frac{a}{2} - \log |v_0| \right). \tag{3.30}$$

Set $\theta(\tau) = |\exp(P_1 h(\tau)) w_0|$; then $\theta(0) = |w_0|$. By (3.27),

$$|v_0| \leq C\theta(0). \tag{3.31}$$

By (3.17),

$$\theta(T) \geq K \exp(\mu h(T)) \theta(0). \tag{3.32}$$

We deduce from (3.29)–(3.32) that

$$\log \left(\frac{2\theta(T)}{a} \right) \geq \log \theta(T) - \log |v_0 \exp(\lambda h(T))| \geq$$

$$\geq \log K + \log \theta(0) - \log |v_0| + (\mu - \lambda) h(T) \geq$$

$$\geq \log K - \log C + \left(\frac{\mu}{\lambda} - 1 \right) \left(\frac{a}{2} - \log |v_0| \right) \geq$$

$$\geq \log K - \log C + \left(\frac{\mu}{\lambda} - 1 \right) \left(\frac{a}{2} - \log b \right) \geq 0.$$

We get a contradiction with (3.28) for $\tau = T$ since the norm of the w-coordinate of $\phi(h(T), s)$ equals $\theta(T)$, while the w-coordinate of $g(T)$ is 0. The lemma is proved. □

Let us complete the proof of Theorem 3.3.1 in case (B1). Assume that $l, \delta_0 > 0$ are chosen for Σ so that the LE holds.

Take $\varepsilon \in (0, \min(\delta_0, \varepsilon_0, a/2))$ so small that if $\text{dist}(y, r) < \varepsilon$, then $\phi(t, y)$ intersects Σ at a point s such that

$$\text{dist}(\phi(t, s), r) < \varepsilon_0, \quad |t| \le l\varepsilon. \tag{3.33}$$

Consider the corresponding d and a d-pseudotrajectory g described above.

Assume that

$$\text{dist}(\phi(h(t), x), g(t)) < \varepsilon, \quad t \in \mathbb{R}, \tag{3.34}$$

for some point x and reparametrization h and set $y = \phi(h(0), x)$.

Then $\text{dist}(y, r) < \varepsilon$, and there exists a point $s = \phi(\tau, y) \in \Sigma$ with $|\tau| < l\varepsilon$.

If $-l\varepsilon \le t \le 0$, then

$$\text{dist}(\phi(t, s), O^-(r, \phi)) \le \varepsilon_0$$

by (3.33).

If $t < -l\varepsilon$, then $h(0) + \tau + t < h(0)$, and there exists $t' < 0$ such that $h(t') = h(0) + \tau + t$. In this case,

$$\phi(t, s) = \phi(h(0) + \tau + t, x) = \phi(h(t'), x),$$

and

$$\text{dist}\,(\phi(t, s), O^-(r, \phi)) \le \text{dist}\left(\phi(h(t'), x), \phi(t', r)\right) \le \varepsilon_0.$$

By Lemma 3.3.2, $s \in r + L$. If ε is small enough, then $\text{dist}(s, r) < b$, where b satisfies the condition of Lemma 3.3.3, whose conclusion contradicts (3.34).

This completes the consideration of case (B1) for $\dim W^u(p) \ge 2$. If $\dim W^u(p) = 1$, then the nontransversality of $W^u(q)$ and $W^s(p)$ implies that $L \subset E^s$. This case is trivial since any shadowing trajectory passing close to r must belong to the intersection $W^u(q) \cap W^s(p)$, while we can construct a pseudotrajectory "going away" from p along $W^u(p)$. If $\dim W^u(p) = 0$, $W^u(q)$ and $W^s(p)$ cannot have a point of nontransverse intersection.

Case (B2) Passing from the vector field X to $-X$, if necessary, we may assume that p is a closed trajectory. We "linearize" X in a neighborhood of p as described in the proof of Lemma 3.3.1 so that the local Poincaré transformation of the transverse section Σ is a linear mapping generated by a matrix P with the following properties: With respect to some coordinates in Σ,

$$P = \text{diag}(P_s, P_u), \tag{3.35}$$

where $|\lambda_j| < 1$ for the eigenvalues λ_j of the matrix P_s, $|\lambda_j| > 1$ for the eigenvalues λ_j of the matrix P_u, every eigenvalue has multiplicity 1, and P is in a Jordan form.

The same reasoning as in case (B1) shows that it is possible to perform such a "linearization" (and other perturbations of X performed below) so that the nontransversality of $W^u(q)$ and $W^s(p)$ is preserved.

Consider an eigenvalue λ of P_u such that $|\lambda| \leq |\mu|$ for the remaining eigenvalues μ of P_u.

We treat separately the following two cases.

Case (B2.1): $\lambda \in \mathbb{R}$.
Case (B2.2): $\lambda \in \mathbb{C} \setminus \mathbb{R}$.

Case (B2.1) Applying a perturbation, we may assume that

$$P_u = \text{diag}(\lambda, P_1),$$

where $|\lambda| < |\mu|$ for the eigenvalues μ of the matrix P_1 (thus, there exists a one-dimensional direction of "weakest expansion" in $W^u(p)$). In this case, we apply precisely the same reasoning as that applied to treat case (B1) (we leave details to the reader).

Case (B2.2) Applying one more perturbation of X, we may assume that

$$\lambda = \nu + i\eta = \rho \exp\left(\frac{2\pi m_1 i}{m}\right),$$

where m_1 and m are relatively prime natural numbers, and

$$P_u = \text{diag}(Q, P_1),$$

where

$$Q = \begin{pmatrix} \nu & -\eta \\ \eta & \nu \end{pmatrix}$$

with respect to some coordinates (y, v, w) in Σ, where $\rho = |\lambda| < |\mu|$ for the eigenvalues μ of the matrix P_1.

Denote

$$E^s = \{(y, 0, 0)\}, \quad E^u = \{(0, v, w)\}, \quad E_1^u = \{(0, v, 0)\}.$$

Thus, E^s is the "stable subspace," E^u is the "unstable subspace," and E_1^u is the two-dimensional "unstable subspace of the weakest expansion."

Geometrically, the Poincaré transformation $\sigma : \Sigma \to \Sigma$ (extended as a linear mapping to E_1^u) acts on E_1^u as follows: the radius of a point is multiplied by ρ, while $2\pi m_1/m$ is added to the polar angle.

As in the proof of Lemma 3.3.1, we take a small neighborhood W of the origin of the transverse section Σ so that, for points $x \in W$, the function $\alpha(x)$ (the time of first return to Σ) is defined.

We assume that the point r of nontransverse intersection of $W^u(q)$ and $W^s(p)$ belongs to the section Σ. Similarly to case (B1), we perturb X so that, in a neighborhood of r, the component of intersection of $W^u(q) \cap \Sigma$ containing r has the form of an affine space, $r + L$.

Let Π^u be the projection in Σ to E^u parallel to E^s and let Π_1^u be the projection to E_1^u; thus,

$$\Pi^u(y, u, v) = (0, u, v) \text{ and } \Pi_1^u(y, u, v) = (0, u, 0).$$

The nontransversality of $W^u(q)$ and $W^s(p)$ at r means that

$$L' = \Pi^u L \neq E^u$$

(see case (B1)). Applying a reasoning similar to that in case (B1), we perturb X so that if $L'' = L' \cap E_1^u$, then

$$\dim L'' < \dim E_1^u = 2.$$

Hence, either $\dim L'' = 1$ or $\dim L'' = 0$. We consider only the first case, the second one is trivial.

Denote by A the line L''. Images of A under degrees of σ (extended to the whole plane E_1^u) are m different lines in E_1^u.

In what follows, we refer to an obvious geometric statement (given without a proof).

Proposition 3.3.1 *Consider coordinates $(x_1, \dots, x_n)$ in the Euclidean space $\mathbb{R}^n$. Let $x' = (x_1, x_2)$, $x'' = (x_3, \dots, x_n)$, and let G be the plane of coordinate x'. Let D be a hyperplane in $\mathbb{R}^n$ such that*

$$D \cap G = \{x_2 = 0\}.$$

For any $b > 0$ there exists $c > 0$ such that if $x = (x', x'') \in D$ and $x' = (x'_1, x'_2)$, then either $|x'_2| \le b|x'_1|$ or $|x''| \ge c|x'|$.

Take $a > 0$ such that the $2a$-neighborhood of the origin in Σ belongs to W. We may assume that if $v = (v_1, v_2)$, then the line A is $\{v_2 = 0\}$.

Take $b > 0$ such that the images of the cone

$$C = \{v : \ |v_2| \le b|v_1|\}$$

in E_1^u under degrees of σ intersect only at the origin (denote these images by $C_1, \dots, C_m$).

We apply Proposition 3.3.1 to find a number $c > 0$ such that if $(0, v, w) \in L'$, then either $(0, v, 0) \in C$ or

$$|w| \geq c|v|. \tag{3.36}$$

Take a point $\beta = (0, v, 0) \in \Sigma$, where $|v| = a$, such that $\beta \notin C_1 \cup \cdots \cup C_m$.

For a natural number N, set $\beta_N = (r_y, P_u^{-N}(v, 0)) \in \Sigma$ (we recall that equality (3.35) holds), where r_y is the y-coordinate of r. We naturally identify β and β_N with points of M and consider the following pseudotrajectory:

$$g(t) = \begin{cases} \phi(t, r), & t \leq 0; \\ \phi(t, \beta_N), & t > 0. \end{cases}$$

The following statement (similar to Lemma 3.3.2) holds: there exists $\varepsilon_0 > 0$ such that if

$$\operatorname{dist}(\phi(t, s), O^-(r, \phi)) < \varepsilon_0, \quad t \leq 0,$$

for some point $s \in \Sigma$, then $s \in r + L$.

Since β does not belong to the closed set $C_1 \cup \cdots \cup C_m$, we may assume that the disk in E_1^u centered at β and having radius ε_0 does not intersect the set $C_1 \cup \cdots \cup C_m$.

Define numbers

$$\alpha_1(N) = \alpha(\beta_N), \ \alpha_2(N) = \alpha_1(N) + \alpha(\sigma(\beta_N)), \ldots,$$

$$\alpha_N(N) = \alpha_{N-1}(N) + \alpha(\sigma^{N-1}(\beta_N)).$$

Take δ_0 and l for which LE holds for the neighborhood W (reducing W, if necessary). Take $\varepsilon < \min(\varepsilon_0/l, \delta_0)$ and assume that there exists the corresponding d (from the definition of the OrientSP$_F$). Take N so large that g is a d-pseudotrajectory.

Let h be a reparametrization; assume that

$$|\phi(h(t), p_0) - g(t)| < \varepsilon, \quad 0 \leq t \leq \alpha_N(N),$$

for some point $p_0 \in \Sigma$.

Since $g(\alpha_k(N)) \in \Sigma$ for $0 \leq k \leq N$ by construction, there exist numbers χ_k such that

$$|\sigma^{\chi_k}(p_0) - g(\alpha_k(N))| < \varepsilon_0, \quad 0 \leq k \leq N.$$

To complete the proof of Theorem 3.3.1, let us show that for any $p_0 \in r + L$ and any reparametrization h there exists $t \in [0, \alpha_N(N)]$ such that

$$\operatorname{dist}(\phi(h(t), p_0), g(t)) \geq \varepsilon.$$

Assuming the contrary, we see that

$$|\sigma^{\chi_k}(p_0) - g(\alpha_k(N))| < \varepsilon_0, \quad 0 \le k \le N,$$

where the numbers χ_k were defined above.

We consider two possible cases.

If

$$\Pi_1^u p_0 \in C$$

(C is the cone defined before estimate (3.36)), then

$$\Pi_1^u \sigma^{\chi_k}(p_0) \in C_1 \cup \cdots \cup C_m.$$

By construction, $\Pi_1^u g(\alpha_N(N))$ is β. Hence,

$$|\Pi_1^u \sigma^{\chi_N}(p_0) - \Pi_1^u g(\alpha_N(N))| > \varepsilon_0,$$

and we get the desired contradiction.

If

$$\Pi_1^u p_0 \notin C$$

and $p_0 = (y_0, v_0, w_0)$, then $(0, v_0, w_0) \in L'$, and it follows from (3.36)) that $|w_0| \ge c|v_0|$. In this case, decreasing ε_0, if necessary, we apply the reasoning similar to Lemma 3.3.3.

Thus, we have proved inclusion (3.9), which completes the proof of Theorem 3.3.1. □

Historical Remarks The first result concerning C^1 interiors of sets of vector fields having some shadowing properties was obtained by K. Lee and the second author in [33]. Denote by $\mathscr{N}$ the set of nonsingular vector fields. It was shown in [33] that vector fields in the set

$$\mathrm{Int}^1(\mathrm{SSP}_F) \cap \mathscr{N}$$

are structurally stable.

The class $\mathscr{B}$ was introduced by S. B. Tikhomirov in [99].

Theorem 3.3.1 was proved by the first author and S. B. Tikhomirov in [69].

Let us also note that S. B. Tikhomirov proved in [99] the following result: If the dimension of the manifold does not exceed 3, then

$$\mathrm{Int}^1(\mathrm{OrientSP}_F) = \mathscr{S}_F.$$

3.4 Vector Fields of the Class $\mathscr{B}$

In the previous section, we defined the set $\mathscr{B}$ of vector fields. As was mentioned, vector fields of that class are not structurally stable. This section is devoted to the following result [69].

Theorem 3.4.1 $Int^1(OrientSP_F) \cap \mathscr{B} \neq \emptyset$.

This theorem states that there exist vector fields in $\text{Int}^1(\text{OrientSP}_F)$ that belong to the class $\mathscr{B}$. The complete proof of Theorem 3.4.1 given in [69] is quite complicated, and we do not give it here.

Instead, we explain the main idea of the proof. One constructs a vector field X of the class $\mathscr{B}$ on the four-dimensional manifold $M = S^2 \times S^2$ that has the following properties (F1)-(F3) (ϕ denotes the flow generated by X).

(F1) The nonwandering set of ϕ is the union of four rest points p, q, s, u.

(F2) We can introduce coordinates in the disjoint neighborhoods $U_p = N(1, p)$ and $U_q = N(1, q)$ so that

$$X(x) = J_p(x - p), \quad x \in U_p,$$

and

$$X(x) = J_q(x - q), \quad x \in U_q,$$

where

$$J_p = \begin{pmatrix} -1 & 0 & 0 & 0 \\ 0 & -2 & 0 & 0 \\ 0 & 0 & 1 & -1 \\ 0 & 0 & 1 & 1 \end{pmatrix}$$

and

$$J_q = \begin{pmatrix} 1 & 0 & 0 & 0 \\ 0 & -1 & 0 & 1 \\ 0 & 0 & 2 & -1 \\ 0 & -1 & 0 & 1 \end{pmatrix}.$$

Since the eigenvalues of J_p are $-1, -2, 1 \pm i$ and the eigenvalues of J_p are $1, 2, -1 \pm i$, conditions (1) and (2) of the definition of the class $\mathscr{B}$ (see the previous section) are satisfied for the vector field X and its rest points q and p.

(F3) The point s is an attracting hyperbolic rest point. The point u is a repelling hyperbolic rest point. The following condition holds:

$$W^u(p) \setminus \{p\} \subset W^s(s), \quad W^s(q) \setminus \{q\} \subset W^u(u). \tag{3.37}$$

The intersection of $W^s(p) \cap W^u(q)$ consists of a single trajectory α, and for any $x \in \alpha$, the condition

$$\dim (T_x W^s(p) \oplus T_x W^u(q)) = 3 \tag{3.38}$$

holds.

These conditions imply that the two-dimensional manifolds $W^s(p)$ and $W^u(q)$ intersect along a one-dimensional curve in the four-dimensional manifold M. Thus, $W^s(p)$ and $W^u(q)$ are not transverse; hence, $X \in \mathscr{B}$.

Geometrically, condition (3.38) means the following. Fix a point $r \in \alpha$ and let Σ be a transverse section to the flow ϕ at r (as above, this means that Σ is a smooth open disk in M of codimension 1 containing r that is transverse to the flow ϕ at any of its points).

Denote by β_s and β_u the intersections of Σ with $W^s(p)$ and $W^u(q)$, respectively. Clearly, β_s and β_u are one-dimensional curves containing the point r. Condition (3.38) means that the curves β_s and β_u intersect at r at nonzero angle.

To prove Theorem 3.4.1, it is enough to show that any vector field X' that is C^1-close to X belongs to OrientSP_F.

The vector field X satisfies Axiom A$'$ and the no-cycle condition; hence, X is Ω-stable. Thus, there exists a neighborhood V of X in $\mathscr{X}^1(M)$ such that for any field $X' \in V$, its nonwandering set consists of four hyperbolic rest points p', q', s', u' that belong to small neighborhoods of p, q, s, u, respectively. We denote by ϕ' the flow of any $X' \in V$ and by $W^s(p')$, $W^u(p')$, etc. the corresponding stable and unstable manifolds.

Select compact subsets b_s and b_u of the curves β_s and β_u, respectively, such that the interiors of b_s and b_u (in the interior topology) contain the point r.

Let Δ_s and Δ_u be compact subsets of $W^s(p)$ and $W^u(q)$, respectively, such that $b_s \subset \Delta_s$ and $b_u \subset \Delta_u$.

It follows from the stable manifold theorem that if $x' \in V$, then the stable and unstable manifolds $W^s(p')$ and $W^u(q')$ of the hyperbolic rest points p' and q' contain compact subsets Δ_s' and Δ_u' that converge (in the C^1 topology) to Δ_s and Δ_u, respectively, as X' tends to X.

Hence, the corresponding curves b_s' and b_u' tend in the C^1 topology to b_s and b_u, respectively, as X' tends to X.

We have the following two possibilities for a vector field $X' \in V$:

- $b_s' \cap b_u' = \emptyset$;
- b_s' and b_u' have a point r' of intersection close to r, and they intersect at r' at nonzero angle.

Clearly, we can choose Σ so that in the first case,

$$W^u(p') \cap W^s(q') = \emptyset;$$

then the vector field X' is structurally stable, and $X' \in \text{OrientSP}_F$.

Thus, it remains to consider the second case. To simplify notation, we write X, ϕ, etc. instead of X', ϕ', etc.

In this case, we make several additional assumptions which help us to explain to the reader the main geometric ideas used in the proof of Theorem 3.4.1 and to avoid heavy technical constructions of [69]. Here we follow the reasoning of [100].

First, we assume that the vector field X is linear in neighborhoods U_p and U_q of the rest points p and q, respectively (see property (F2) above).

In addition, we assume that, in a sense, the shift at some fixed time along trajectories in a neighborhood of a compact part of the trajectory α of nontransverse intersection of $W^s(p)$ and $W^u(q)$ is a parallel translation (see property (F5) below).

Let us introduce some notation. For a point $x \in U_p$ denote $P_1x = x_1$ and $P_{34}x = (x_3, x_4)$, where $x - p = (x_1, x_2, x_3, x_4)$; for a point $x \in U_q$, denote $P_1x = x_1$ and $P_{24}x = (x_2, x_4)$, where $x - q = (x_1, x_2, x_3, x_4)$. For a small $m > 0$ we denote $W^u_{loc}(p, m) = W^u(p) \cap N(m, p)$ etc.

Our additional assumptions are as follows.

(F4) The trajectory α satisfies the following inclusions:

$$\alpha \cap U_p \subset \{p + (t, 0, 0, 0); t \in (0, 1)\} \text{ and } \alpha \cap U_q \subset \{q - (t, 0, 0, 0); t \in (0, 1)\}.$$

(F5) There exist numbers $\Delta \in (0, 1)$ and $T_a > 0$ such that

$$\phi(T_a, q + (-1, x_2, x_3, x_4)) = (p + (1, x_2, x_3, x_4)), \quad |x_2|, |x_3|, |x_4| < \Delta.$$

(F6) $\phi(t, x) \notin U_q$ for $x \in U_p$, $t \geq 0$.

In what follows, we need two simple geometric lemmas.

In the first lemma, we consider a planar linear system of differential equations

$$\frac{dx}{dt} = Jx, \quad x \in \mathbb{R}^2,$$

where

$$J = \begin{pmatrix} 1 & -1 \\ 1 & 1 \end{pmatrix},$$

and denote by $\psi(t, x)$ its flow on $\mathbb{R}^2$.

If a point $x \in \mathbb{R}^2$ has polar coordinates (r, θ) with $\theta \in [0, 2\pi)$ and $r \neq 0$, we put $\arg(x) = \theta$.

Lemma 3.4.1 *For any point $x_0 \in \mathbb{R}^2 \setminus 0$, angle $\Theta \in [0, 2\pi)$, and number T_0 there exists $t < T_0$ such that $\arg(\psi(t, x_0)) = \Theta$.*

The proof of this lemma is straightforward. Of course, a similar statement holds for the system

$$\frac{dx}{dt} = -Jx, \quad x \in \mathbb{R}^2,$$

with $t < T_0$ replaced by $t > T_0$.

Lemma 3.4.2 *Let S_1 and S_2 be three-dimensional vector spaces with coordinates (x_1, x_2, x_3) and (y_1, y_2, y_3), respectively. Let $Q : S_2 \to S_1$ be a linear map such that*

$$Q\{y_2 = y_3 = 0\} \neq \{x_2 = x_3 = 0\}.$$

Then for any $D > 0$ there exists $R > 0$ (depending on Q and D) such that if two sets $V_1 \subset S_1 \cap \{x_1 = 0\}$ and $V_2 \subset S_2 \cap \{y_1 = 0\}$ satisfy the following conditions:

- *$V_1 \subset N(R, 0)$ and $V_2 \subset N(R, 0)$;*
- *V_1 intersects any ray in $S_1 \cap \{x_1 = 0\}$ starting at 0;*
- *V_2 intersects any ray in $S_2 \cap \{y_1 = 0\}$ starting at 0;*

then

$$C_1 \cap QC_2 \neq \emptyset,$$

where

$$C_1 = \{(x_1, x_2, x_3) : \quad |x_1| < D, \ (0, x_2, x_3) \in V_1\}$$

and

$$C_2 = \{(y_1, y_2, y_3) : \quad |y_1| < D, \ (0, y_2, y_3) \in V_2\}.$$

Proof Let us fix a linear map Q and a number $D > 0$. Consider the lines $l_1 \subset S_1$ and $l_2 \subset S_2$ given by the equations $x_2 = x_3 = 0$ and $y_2 = y_3 = 0$, respectively.

By our assumption, $Ql_2 \neq l_1$. Let us consider the plane $\pi \subset S_1$ containing l_1 and Ql_2. Consider a parallelogram $P \subset \pi$ that is symmetric with respect to 0, has sides parallel to l_1 and Ql_2, and satisfies the relation

$$P \subset \{|x_1| < D\} \cap Q(\{|y_1| < D\}). \tag{3.39}$$

Find a number $R > 0$ such that the following inclusions hold:

$$B(R, 0) \cap \pi \subset P \quad \text{and} \quad Q(B(R, 0) \cap Q^{-1}\pi) \subset P. \tag{3.40}$$

Let z_1 be a point of intersection of V_1 and the line $\pi \cap \{x_1 = 0\}$. Condition (3.40) implies that $z_1 \in P$. Consider the line k_1 containing z_1 and parallel to l_1. Inclusion (3.39) implies that $k_1 \cap P \subset C_1$.

Similarly, let z_2 be a point of intersection of V_2 and the line $\pi \cap \{y_1 = 0\}$. Condition (3.40) implies the inclusion $Qz_2 \in P$. Let k_2 be the line containing Qz_2 and parallel to Ql_2. Inclusion (3.39) implies that $Q^{-1}(k_2 \cap V) \subset C_2$.

Since $k_1 \nparallel k_2$, there exists a point $z \in k_1 \cap k_2$. The inclusions $z_1, z_2 \in P$ imply that $z \in P$. Hence, $z \in C_1 \cap QC_2$. Our lemma is proved. □

Now let us prove that the vector field X has the oriented shadowing property.

Fix points $y_p = \alpha(T_p) \in U_p$ and $y_q = \alpha(T_q) \in U_q$ (note that in this case, $T_p > T_q$ by property (F5)) and a number $\delta > 0$.

We say that $g(t)$ is a pseudotrajectory of type $\mathrm{Ps}(\delta)$ if

$$g(t) = \begin{cases} \phi(t - T_p, x_p), & t > T_p; \\ \phi(t - T_q, x_q), & t < T_q; \\ \alpha(t), & t \in [T_q, T_p], \end{cases}$$

for some points $x_p \in B(\delta, y_p)$ and $x_q \in B(\delta, y_q)$.

Fix an $\varepsilon > 0$. Let us say that a pseudotrajectory $g(t)$ can be ε-oriented shadowed if there exists a reparametrization $h \in \mathrm{Rep}$ and a point z such that

$$\mathrm{dist}(\phi(h(t), z), g(t)) < \varepsilon, \quad t \in \mathbb{R}.$$

Clearly, the required inclusion $X \in \mathrm{OrientSP}_F$ is a corollary of the following two statements.

Proposition 3.4.1 *For any $\delta > 0$, $y_p \in \alpha \cap U_p$, and $y_q \in \alpha \cap U_q$ there exists $d > 0$ such that if $g(t)$ is a d-pseudotrajectory of X, then either $g(t)$ can be ε-oriented shadowed or there exists a pseudotrajectory $g^*(t)$ of type $Ps(\delta)$ with these y_p and y_q and a number $t_0 \in \mathbb{R}$ such that*

$$dist\left(g(t), g^*(t + t_0)\right) < \varepsilon/2, \quad t \in \mathbb{R}.$$

Proposition 3.4.2 *There exist $\delta > 0$, $y_p \in \alpha \cap U_p$, and $y_q \in \alpha \cap U_q$ such that any pseudotrajectory of type $Ps(\delta)$ with these y_p and y_q can be $\varepsilon/2$-oriented shadowed.*

Proposition 3.4.1 can be proved by a standard reasoning. Precisely the same statement was proved in [69] for a slightly different vector field (the only difference is in the structure of the matrices J_p and J_q). The proof can be literally repeated in our case.

The main idea of the proof is the following. Outside a neighborhood of the curve α, our vector field X coincides with a structurally stable one. Hence, pseudotrajectories that do not intersect a fixed neighborhood of α can be shadowed.

If $g(t)$ intersects a small neighborhood of α, then (after a proper shift of time), the points $g(t)$ with $t > T_p$ also belong to a set where X coincides with a structurally stable vector field; thus, for such t, $g(t)$ can be shadowed by $\phi(t - T_p, x_p)$. Similarly, the pseudotrajectory $g(t)$ can be shadowed by $\phi(t - T_q, x_q)$. For $t \in (T_q, T_p)$, the points $g(t)$ are close to α. We leave the rest of the proof to the reader.

Proof (of Proposition 3.4.2) Since the rest points s and u are a hyperbolic attractor and a hyperbolic repeller, we may assume, without loss of generality, that

$$O^+(N(\varepsilon/2,s),\phi) \subset N(\varepsilon,s) \quad \text{and} \quad O^-(N(\varepsilon/2,u),\phi) \subset N(\varepsilon,u),$$

where $O^+(A,\phi)$ and $O^-(A,\phi)$ are the positive and negative semitrajectories of a set A in the flow ϕ, respectively.

Take $m \in (0,\varepsilon/8)$. We fix points $y_p = \alpha(T_p) \in N(m/2,p) \cap \alpha$ and $y_q = \alpha(T_q) \in N(m/2,q) \cap \alpha$. Put $T = T_p - T_q$. Find a number $\delta > 0$ such that if $g(t)$ is a pseudotrajectory of type $\mathrm{Ps}(\delta)$ (with y_p and y_q fixed above), $t_0 \in \mathbb{R}$, and $x_0 \in N(2\delta, g(t_0))$, then

$$\mathrm{dist}(\phi(t-t_0,x_0), g(t)) < \varepsilon/2, \quad |t-t_0| \le T+1. \tag{3.41}$$

Consider a number $\tau > 0$ such that if $x \in W^u(p) \setminus N(m/2,p)$, then $\phi(\tau,x) \in N(\varepsilon/8,s)$. Take $\varepsilon_1 \in (0,m/4)$ such that if two points $z_1,z_2 \in M$ satisfy the inequality $\mathrm{dist}(z_1,z_2) < \varepsilon_1$, then

$$\mathrm{dist}(\phi(t,z_1),\phi(t,z_2)) < \varepsilon/8, \quad |t| \le \tau.$$

In this case, for any $y \in N(\varepsilon_1,x)$, the following inequalities hold:

$$\mathrm{dist}(\phi(t,x),\phi(t,y)) < \varepsilon/4, \quad t \ge 0. \tag{3.42}$$

Decreasing ε_1, we may assume that if $x' \in W^s(q) \setminus N(m/2,q)$ and $y' \in N(\varepsilon_1,x')$, then

$$\mathrm{dist}(\phi(t,x'),\phi(t,y')) < \varepsilon/4, \quad t \le 0.$$

Let $g(t)$ be a pseudotrajectory of type $\mathrm{Ps}(\delta)$, where y_p, y_q, and δ satisfy the above-formulated conditions.

Let us consider several possible cases. t

Case (P1): $x_p \notin W^s(p)$ and $x_q \notin W^u(q)$. Let

$$T' = \inf\{t \in \mathbb{R} : \phi(t,x_p) \notin N(p,3m/4)\}.$$

If δ is small enough, then $\mathrm{dist}(\phi(T',x_p), W^u(p)) < \varepsilon_1$. In this case, there exists a point $z_p \in W^u_{loc}(p,m) \setminus N(m/2,p)$ such that

$$\mathrm{dist}(\phi(T',x_p),z_p) < \varepsilon_1. \tag{3.43}$$

Applying a similar reasoning in a neighborhood of q (and reducing δ, if necessary), we find a point $z_q \in W^s_{loc}(q,m) \setminus N(m/2,q)$ and a number $T'' < 0$ such that $\mathrm{dist}(\phi(T'',x_q),z_q) < \varepsilon_1$.

Consider the hyperplanes $S_p := \{x_1 = P_1 y_p\}$ and $S_q := \{x_1 = P_1 y_q\}$. From our assumptions on the linearity of X in neighborhoods of p and q and from assumption (F5) it follows that the Poincaré map defined by $Q(x) = \phi(T, x)$ is a linear map $Q : S_q \to S_p$ such that $Q(\{(x_2, x_4) = 0\}) \neq \{(x_3, x_4) = 0\}$.

Apply Lemma 3.4.2 to the hyperplanes S_p and S_q, the map Q, and the number $D = \varepsilon/8$ and find the corresponding $R > 0$. Note that there exists a $T_R > 0$ such that

$$|\phi(t, P_{34}x_p)| < R,\ t < -T_R, \text{ and } |\phi(t, P_{24}x_q)| < R,\ t > T_R.$$

Consider the sets

$$V^- = \{\phi(t, P_{34}x_p) : t < -T_R\} \text{ and } V^+ = \{\phi(t, P_{24}x_q) : t > T_R\}.$$

Due to Lemma 3.4.1, the sets $V^\pm$ satisfy the assumptions of Lemma 3.4.2; hence, the sets

$$C^- = \{x \in S_p : \quad P_{34}x \in V^-, |P_2 x| < D\}$$

and

$$C^+ = \{x \in S_q : \quad P_{24}x \in V^+, |P_3 x| < D\}$$

are such that $C^- \cap QC^+ \neq \emptyset$.

Let us consider a point

$$x_0 \in C^- \cap QC^+ \tag{3.44}$$

and numbers $t_p < -T_R$ and $t_q > T_R$ such that $P_{34}x_0 = \phi(t_p, P_{34}x_s)$ and $P_{24}Q^{-1}x_0 = \phi(t_q, P_{24}x_u)$. The following inclusions hold:

$$\phi(-T_Q - T_R - T'', x_0) \in N(2\varepsilon_1, z_q), \quad \phi(-T_Q, x_0) \in N(D, y_q),$$

$$\phi(0, x_0) \in N(D, y_p), \quad \phi(T_R + T', x_0) \in N(2\varepsilon_1, z_p).$$

Inequalities (3.41) imply that if δ is small enough, then

$$\operatorname{dist}(\phi(t_3 + t, x_0), g(T_p + t)) < \varepsilon/2, \quad t \in [-T, 0]. \tag{3.45}$$

Define a reparametrization $h(t)$ as follows:

$$h(t) = \begin{cases} h(T_q + T'' + t) = -T_Q - T_R - T'' + t, & t < 0; \\ h(T_p + T' + t) = T_R + T' + t, & t > 0; \\ h(T_p + t) = t, & t \in [-T, 0]; \\ h(t) \text{ increases}, & t \in [T_p, T_p + T'] \cup [T_q + T'', T_q]. \end{cases}$$

If $t \geq T_p + T'$, then inequality (3.42) implies that

$$\operatorname{dist}(\phi(h(t), x_0), \phi(t - (T_p + T'), z_p)) < \varepsilon/4$$

and

$$\operatorname{dist}(\phi(t - T_p, x_p), \phi(t - (T_p + T'), z_p)) < \varepsilon/4.$$

Hence, if $t \geq T_p + T'$, then

$$\operatorname{dist}(\phi(h(t), x_0), g(t)) < \varepsilon/2. \tag{3.46}$$

For $t \in [T_p, T_p + T']$, the inclusions $\phi(h(t), x_0), g(t) \in N(m, p)$ hold, and inequality (3.46) holds for these t as well.

A similar reasoning shows that inequality (3.46) holds for $t \leq T_q$. If $t \in [T_q, T_p]$, then inequality (3.46) follows from (3.45). This completes the proof in case (P1).

Case (P2): $x_p \in W^s(p)$ and $x_q \notin W^u(q)$. In this case, the proof uses the same reasoning as in case (P1). The only difference is that instead of (3.44) we construct a point $x_0 \in N(D, y_p) \cap W^s_{loc}(p, m)$ such that

$$\phi(-T - T'', x_0) \in N(2\varepsilon_1, z_q) \text{ and } \phi(-T, x_0) \in N(\varepsilon/8, y_q).$$

The construction is straightforward and uses Lemma 3.4.1.

Case (P3): $x_p \notin W^s(p)$ and $x_q \in W^u(q)$. This case is similar to case (P2).

Case (P4): $x_p \in W^s(p)$ and $x_q \in W^u(q)$. In this case, we take α as the shadowing trajectory; the reparametrization is constructed similarly to case (P1).

Thus, we have shown that $X \in \text{OrientSP}_F$. □

Historical Remarks Theorem 3.4.1 was published by the first author and S. B. Tikhomirov in [69]. As was said at the beginning of Chap. 3, the complete proof given in this paper is technically very complicated, and we only describe a "model" published by S. B. Tikhomirov in the paper [100] devoted to the Komuro conjecture [29].

Chapter 4
Chain Transitive Sets and Shadowing

In this chapter, we study relations between the shadowing property of diffeomorphisms on their chain transitive sets and the hyperbolicity of such sets.

We prove the following two main results:

- Let Λ be a closed invariant set of $f \in \mathrm{Diff}^1(M)$. Then $f|_\Lambda$ is chain transitive and C^1-stably shadowing in a neighborhood of Λ if and only if Λ is a hyperbolic basic set (Theorem 4.2.1);
- there is a residual set $\mathscr{R} \subset \mathrm{Diff}^1(M)$ such that if $f \in \mathscr{R}$ and Λ is a locally maximal chain transitive set of f, then Λ is hyperbolic if and only if $f|_\Lambda$ is shadowing (Theorem 4.3.1).

The structure of the chapter is as follows.

In Sect. 4.1, we discuss several examples of chain transitive sets. Section 4.2 is devoted to the proof of Theorem 4.2.1. In Sect. 4.3, we prove Theorem 4.3.1.

4.1 Examples of Chain Transitive Sets (Homoclinic Classes)

Let M be a closed smooth manifold and let, as above, $\mathrm{Diff}^1(M)$ be the space of diffeomorphisms of M with the C^1 topology.

Consider a diffeomorphism $f \in \mathrm{Diff}^1(M)$ and its invariant set A. Denote by $f_{|A}$ the restriction of f to A.

Let $\Lambda \subset M$ be a closed f-invariant set. We say that $f|_\Lambda$ has the *standard shadowing property* if for any $\varepsilon > 0$ there is $d > 0$ such that for any d-pseudotrajectory $\{x_i\}_{i=a}^{b} \subset \Lambda$ of f, where $-\infty \le a < b \le \infty$, there is a point $y \in M$ such that

$$\mathrm{dist}\left(f^i(y), x_i\right) < \varepsilon, \quad a \le i \le b-1.$$

S.Yu. Pilyugin, K. Sakai, *Shadowing and Hyperbolicity*, Lecture Notes in Mathematics 2193, DOI 10.1007/978-3-319-65184-2_4

In what follows, in this chapter we write just "shadowing property" instead of "standard shadowing property."

Notice that we consider d-pseudotrajectories of f "contained in Λ," while the shadowing points $y \in M$ are not necessarily contained in Λ.

Let

$$\mathscr{R}(f) = \{x \in M : x \leftrightsquigarrow x\}$$

be the chain recurrent set of f (see Chap. 1).

Then

$$\mathrm{Per}(f) \subset \Omega(f) \subset \mathscr{R}(f).$$

It is easy to see that if f has the shadowing property (on M), then $\Omega(f) = \mathscr{R}(f)$.

The relation $\leftrightsquigarrow$ induces on $\mathscr{R}(f)$ an equivalence relation, whose equivalence classes are called chain recurrence classes of f.

Recall that a closed f-invariant set Λ is called *chain transitive* if

$$x \leftrightsquigarrow_{\Lambda} y$$

for any $x, y \in \Lambda$. It is known that every chain recurrence class is chain transitive (see Proposition 1.1.1).

Let $p, q \in \mathrm{Per}(f)$ be hyperbolic saddle periodic points of f.

We say that p and q are *homoclinically related* and write $p \sim q$ if either $W^s(p)$ and $W^u(q)$ or $W^u(p)$ and $W^s(q)$ have points of transverse intersection.

Let $H_f(p)$ be the *homoclinic class* of p, i.e., the closure of the set of all $q \in \mathrm{Per}(f)$ such that $p \sim q$.

Note that, by the Smale's transverse homoclinic point theorem, $H_f(p)$ coincides with the closure of the set of transverse homoclinic points $x \in W^s(p) \cap W^u(p)$.

The following version of Smale's theorem is stated in [96].

Theorem 4.1.1 *Let p be a hyperbolic periodic point of the diffeomorphism f and let x be a transverse homoclinic point of p. Then every neighborhood of x contains infinitely many periodic points that are homoclinically related to p.*

If a homoclinic class is not hyperbolic, it may contain periodic points having different indices, i.e., there may exist periodic points q_1 and q_2 in the class such that $\dim E^s(q_1) \neq \dim E^s(q_2)$.

In fact, there are examples of diffeomorphisms with homoclinic classes containing hyperbolic periodic points with different indices and such that this phenomenon is preserved under C^1-small perturbations (see, for example, [11]).

Let, as above, p be a hyperbolic periodic point of the diffeomorphism f and set

$$H_f(O(p,f)) = H_f(p) \cup \cdots \cup H_f(f^{\pi(p)-1}(p))$$

(recall that $\pi(p)$ is the minimal period of a periodic point p).

Denote by $C_f(p)$ the chain recurrence class containing p. Then $H_f(O(p,f)) \subset C_f(p)$, but, in general, these sets do not coincide.

Obviously, $H_f(O(p,f))$ is a closed f-invariant set, and it is known that $f|_{H_f(O(p,f))}$ is transitive (see [48]).

4.1.1 Chain Transitive Sets Without Periodic Points

Note that chain transitive sets do not necessarily contain periodic points.

It is well known that every irrational rotation map on the unit circle S^1 is transitive but does not have periodic points.

More generally, there is a translation of the n-dimensional torus $\mathbb{T}^n$ with the same property.

In the case $n = 2$, let

$$L : \mathbb{T}^2 \to \mathbb{T}^2$$

be a translation defined by

$$(x, y) \mapsto (x + a, y + b),$$

where $(a, b) \in \mathbb{T}^2$ satisfy the property that $va + wb \notin \mathbb{Z}$ for any pair $(v, w) \in \mathbb{Z}^2$ (for instance, if $a = \sqrt{2}/2$ and $b = \sqrt{3}/2$, then $va + wb \notin \mathbb{Z}$ for any pair $(v, w) \in \mathbb{Z}^2$).

Then it follows from [43, Proposition 2.7 and Theorem 3.2] that L is minimal (and hence, it is transitive), but L does not have periodic points. It is not hard to show that L does not have the shadowing property.

4.1.2 Hyperbolic Horseshoes

Smale's hyperbolic horseshoe Λ_f on the two-dimensional disk containing a saddle fixed point p is a typical example of a chain transitive set such that $\Lambda_f = C_f(p)$.

More precisely, let $D^2 \subset \mathbb{R}^2$ be a two-dimensional disk and let f be Smale's horseshoe map on D^2 with a hyperbolic saddle fixed point p (see Fig. 4.1(a)). Denote by Λ_f the hyperbolic horseshoe (containing p) and by $C_f(p)$ the chain recurrence class containing p.

Then $\Lambda_f = H_f(p) = C_f(p)$. Since Λ_f is hyperbolic, $f|_{\Lambda_f}$ has the shadowing property.

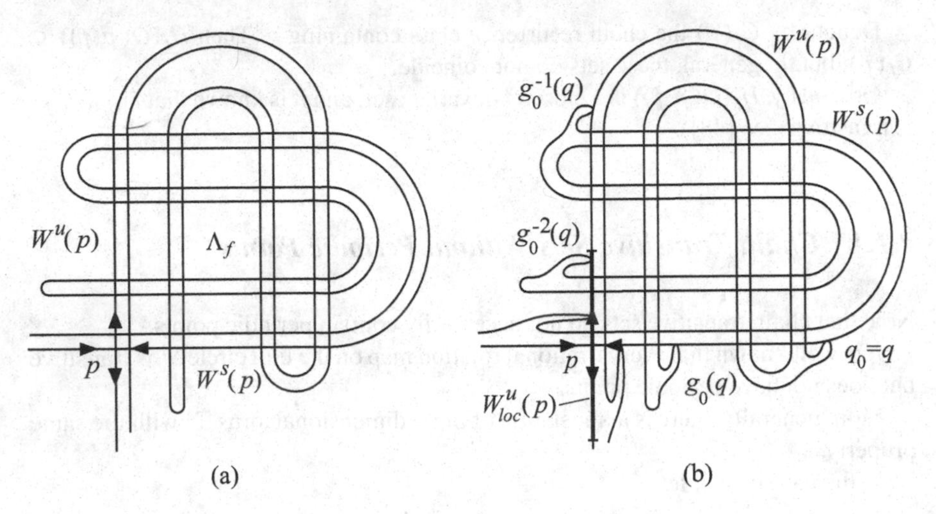

Fig. 4.1 Horseshoes

4.1.3 Horseshoe with a Homoclinic Tangency

Let $\{g_t\}_{t\in\mathbb{R}}$ be the bifurcating one-parameter family of diffeomorphisms on D^2 derived from the horseshoe Λ_f and exhibiting a homoclinic tangency q_0 of g_0 associated to the fixed point p (see [54, Chap. 5]).

Then $\Lambda_f = H_{g_0}(p)$ and $C_{g_0}(p) = \Lambda_f \cup O(q_0, g_0)$ (see Fig. 4.1(b)). Thus, homoclinic classes are not necessarily chain recurrence classes.

We can show that $g_0|_{C_{g_0}(p)}$ does not have the shadowing property. To prove this, for the sake of simplicity, denote g_0 and q_0 by g and q, respectively.

Let $W^u_{loc}(p)$ be a local unstable manifold of p and fix $k > 0$ such that $g^{-k}(q)$ is in the interior of $g(W^u_{loc}(p)) \setminus W^u_{loc}(p)$.

For simplicity, we assume that $k = 1$ (see Fig. 4.2(a)).

Take $\varepsilon > 0$ and denote by $C^u_\varepsilon(q)$ the connected component of $g^2(W^u_{loc}(p)) \cap N(\varepsilon, q)$ containing q.

It is easy to see that there exists $\varepsilon_1 > 0$ such that

$$C^u_{\varepsilon_1}(q) = g^2(W^u_{loc}(p)) \cap N(\varepsilon_1, q).$$

Further, we may assume that if $|g^{-n}(q) - g^{-n}(y)| < \varepsilon_1$ for all $n \geq 0$, then $g^{-2}(y) \in W^u_{loc}(p)$.

To get a contradiction, assume that $g|_{C_g(p)}$ has the shadowing property and let $d = d(\epsilon_1) > 0$ be the corresponding number given by the shadowing property.

Take $l > 0$ such that $|g^l(q) - g^{-l}(q)| < d$. Then

$$\{\ldots, g^{-2}(q), g^{-1}(q), q, g(q), \ldots, g^{l-1}(q), g^{-l}(q), g^{-l+1}(q), \ldots\} \subset C_g(p)$$

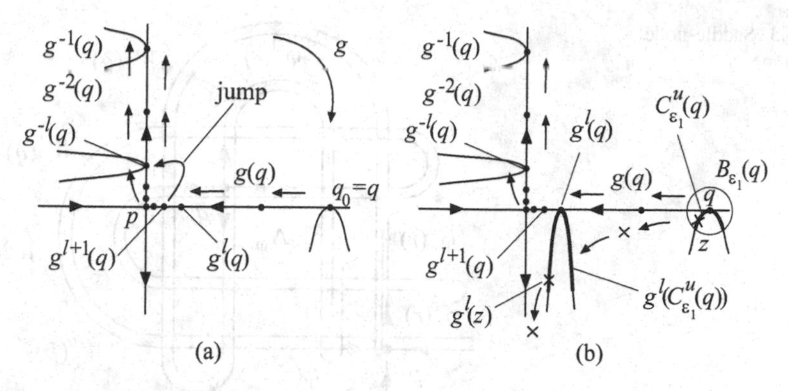

Fig. 4.2 A pseudotrajectory which cannot be shadowable

is a d-pseudotrajectory of g composed of two segments of true g orbits (see Fig. 4.2(a)).

Since $g|_{C_g(p)}$ has the shadowing property, there is a point z close to q that ε_1-shadows the above pseudotrajectory. Thus,

$$\max\left\{\left|g^{-n}(q)-g^{-n}(z)\right|,\ \left|g^{-l+n}(q)-g^{l+n}(z)\right|\right\}<\varepsilon_1 \text{ for all } n\geq 0. \qquad (*)$$

From here it follows that $g^{-2}(z)\in W^u_{loc}(p)$, so that $z\in C^u_{\epsilon_1}(q)$ by the choice of ε_1.

If $z=q$, then the forward orbit of $g^l(z)$ cannot ε_1-shadow the forward orbit of $g^{-l}(q)$ because the ω-limit set of

$$O(g^l(z),g)=O(g^l(q),g)$$

is p. If $z\neq q$, then the forward orbits of $g^l(z)$ and $g^{-l}(q)$ move in opposite directions since p is hyperbolic (see Fig. 4.2(b)).This contradicts $(*)$, and the assertion is proved (for more information, see [90, Sec. 2, 2.2]).

4.1.4 Critical Saddle-Node Horseshoe

Let $\varphi: D^2\to D^2$ be the saddle-node horseshoe map constructed in [51, Sec. 2, 2.2] and possessing a saddle-node fixed point $\tilde{p}$ (see Fig. 4.3).

It is stated in [51] that the saddle-node horseshoe Λ_φ is conjugate to Smale's horseshoe Λ_f. Note that there is a hyperbolic saddle fixed point $\tilde{q}$ in Λ_φ having negative stable and unstable eigenvalues.

We consider here the chain recurrence class $C_\varphi(\tilde{q})$ containing $\tilde{q}$. It is easy to see that

$$\Lambda_\varphi\ (=H_\varphi(\tilde{q}))\subsetneqq C_\varphi(\tilde{q}).$$

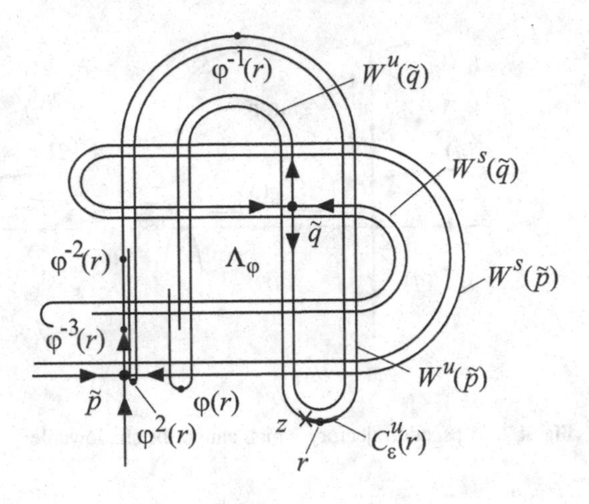

Fig. 4.3 Saddle-node horseshoe

Indeed, if we take $r \in W^u(\tilde{p})$ as in Fig. 4.3, then $r \in C_\varphi(\tilde{q}) \setminus \Lambda_\varphi$ since $\varphi^{\pm n}(r) \to \tilde{p}$ as $n \to \infty$.

We show that, due to the existence of a saddle-node fixed point $\tilde{p}$, $\varphi|_{C_\varphi(\tilde{q})}$ does not have the shadowing property.

To show this, let $r \in W^u(\tilde{p})$ be as above. To get a contradiction, assume that $\varphi|_{C_\varphi(\tilde{q})}$ has the shadowing property. Fix $\varepsilon > 0$ small enough and denote by $C^u_\varepsilon(r)$ the connected component of $W^u(\tilde{p})$ containing r (defined as in the previous example in Sect. 4.1.2, see Fig. 4.3). Let $d = d(\varepsilon) > 0$ be the number corresponding to ε due to the shadowing property of $\varphi|_{C_\varphi(\tilde{q})}$. Take $l > 0$ such that $|\varphi^{-l}(r) - \varphi^{l}(r)| < d$. Then the union of two segments of true φ-orbits,

$$\{\dots, \varphi^{-2}(r), \varphi^{-1}(r), r, \varphi(r), \dots, \varphi^{l-1}(r), \varphi^{-l}(r), \varphi^{-l+1}(r), \dots\} \subset C_\varphi(\tilde{q}),$$

is a d-pseudotrajectory of φ.

Then the same reasoning as in Sect. 4.1.2 shows that there is point $z \in C^u_\varepsilon(r)$ (near r) that ε-shadows the above pseudotrajectory.

Note that $\varphi^n(z) \to \tilde{p}$ as $n \to \infty$ since $\tilde{p}$ is a saddle-node point. Thus, the forward φ-orbit of z cannot ε-shadow the forward φ-orbit of $\varphi^{-l}(r)$. This is the required contradiction (for more information, see [90, Sec. 2, 2.3]).

Historical Remarks Homoclinic orbits and the associated complexity were discovered by H. Poincaré around 1890 (see [75]).

Seventy years after Poincaré, S. Smale constructed in [96, 97] a very simple geometric example (horseshoe) which helped to completely analyze all the complexity found before.

This was the beginning of the geometric theory which we now know as hyperbolic dynamics. The history and many examples involving homoclinic orbits are well described in [54].

4.2 C^1-Stably Shadowing Chain Transitive Sets

In the previous section, we have defined the (standard) shadowing property of a diffeomorphism $f \in \text{Diff}^1(M)$ on a closed f-invariant set $\Lambda \subset M$.

Clearly, this property does not depend on the metric used and is preserved under topological conjugacy. In addition, $f_{|\Lambda}$ has the shadowing property if and only if $f^n_{|\Lambda}$ has the shadowing property for every $n \in \mathbb{Z}\setminus\{0\}$.

Let U be a compact subset of M and put

$$\Lambda_f(U) = \bigcap_{n\in\mathbb{Z}} f^n(U).$$

We say that $f_{|\Lambda}$, or simply Λ, is *locally maximal* (*in* U) if there is a compact neighborhood U of Λ such that $\Lambda = \Lambda_f(U)$. Such a set U is called an *isolating block*. Note that if Λ is locally maximal in U, then $\Lambda = \Lambda_f(U) = \Lambda_f(V)$ for any compact neighborhood $V \subset U$ of Λ.

Denote by $\text{Int}^1(\text{SSP}_D(U))$ the C^1 interior of the set of diffeomorphisms $f \in \text{Diff}^1(f)$ such that $f|_{\Lambda_f(U)}$ is shadowing. Clearly, if $\Lambda = M$, then

$$\text{Int}^1(\text{SSP}_D(M)) = \text{Int}^1(\text{SSP}_D).$$

Thus, $f \in \text{Int}^1(\text{SSP}_D(U))$ if and only if there is a C^1 neighborhood $\mathscr{U}(f)$ of f such that for any $g \in \mathscr{U}(f)$, $g|_{\Lambda_g(U)}$ is shadowing. The set

$$\Lambda_g(U) = \bigcap_{n\in\mathbb{Z}} g^n(U)$$

is called the *continuation* of $\Lambda_f(U)$. We say that $f|_\Lambda$ is *C^1-stably shadowing* (*in* U) if Λ is locally maximal in U and $f \in \text{Int}^1(\text{SSP}_D(U))$.

It is well known that if Λ is hyperbolic, then $f|_\Lambda$ is shadowing (see Theorem 1.4.2). We say that Λ is a *basic set* if Λ is locally maximal and $f|_\Lambda$ is transitive. It is well known that periodic points are dense in hyperbolic basic sets (see [84]).

In this section, we prove the following main result.

Theorem 4.2.1 *Let Λ be a closed invariant set of $f \in \text{Diff}^1(M)$. Then $f|_\Lambda$ is chain transitive and C^1-stably shadowing in U if and only if Λ is a hyperbolic basic set.*

Proof The proof of the "if" statement in Theorem 4.2.1 is easy. Indeed, if Λ is a hyperbolic basic set of f, then $f|_\Lambda$ is transitive (thus, chain transitive) and locally maximal by definition. Let U be a compact neighborhood of Λ in which Λ is locally maximal. Then $f \in \text{Int}^1(\text{SSP}_D(U))$ by the local stability of hyperbolic basic sets [84] since $f|_{\Lambda_f(U)} = f|_\Lambda$ is shadowing.

Let $p \in \text{Per}(f)$ be a hyperbolic saddle periodic point of f with minimal period $\pi(p) > 0$. Recall that $H_f(p)$ is the homoclinic class of p, i.e., the closure of the set of all transverse intersection points $x \in W^s(p) \cap W^u(p)$. Set

$$H_f(O(p,f)) = H_f(p) \cup \cdots \cup H_f\left(f^{\pi(p)-1}(p)\right).$$

Obviously, $H_f(O(p,f))$ is a closed f-invariant set.

Let $\mathscr{C}_f$ be a chain recurrence class of f (recall that $\mathscr{C}_f$ is chain transitive) and assume that $f|_{\mathscr{C}_f}$ is C^1-stably shadowing. Then, by Theorem 4.2.1, there is a saddle $p \in \mathscr{C}_f \cap \mathrm{Per}(f)$ since $\mathscr{C}_f$ is a hyperbolic basic set. From this it follows that $\mathscr{C}_f \subset H_f(O(p,f))$ (see Lemma 4.2.7). On the other hand, it is not difficult to show that every hyperbolic chain recurrence class is locally maximal (Lemma 4.2.8). Thus, we can obtain the next result.

Corollary 4.2.1 *Let $\mathscr{C}_f$ be A chain recurrence class of f. Then $f|_{\mathscr{C}_f}$ is C^1-stably shadowing if and only if there is a hyperbolic saddle $p \in \mathscr{C}_f \cap Per(f)$ such that $\mathscr{C}_f = H_f(O(p,f))$, and $\mathscr{C}_f$ is hyperbolic.*

Now we turn to the proof of the "only if" statement in Theorem 4.2.1.

4.2.1 Preliminaries

Let $f \in \mathrm{Diff}^1(M)$. Throughout this subsection, let Λ be a (nontrivial) closed f-invariant set.

Recall that $f|_\Lambda$ is transitive if there is a point $x \in \Lambda$ such that the omega-limit set $\omega_f(x)$ of x coincides with Λ. Obviously, the notion of chain transitivity is a strict generalization of that of transitivity. The proof of the following lemma is simple and left to the reader.

Lemma 4.2.1 *Assume that $f|_\Lambda$ is locally maximal in U and shadowing. Then*

- *for any pseudotrajectory of f in Λ, the shadowing point can be taken from Λ;*
- *if $f|_\Lambda$ is chain transitive, then $f|_\Lambda$ is transitive.*

The proof of the following lemma is almost the same as that of Lemma 3.1.2.

Lemma 4.2.2 *Assume that $f|_\Lambda$ is C^1-stably shadowing in U. Then there exists a neighborhood $\mathscr{U}(f)$ such that for any $g \in \mathscr{U}(f)$, every point $q \in \Lambda_g(U) \cap Per(g)$ is hyperbolic.*

Before describing our technical results, we have to prepare some notation.

Recall (see Definition 1.3.12) that Λ admits a dominated splitting if the tangent bundle $T_\Lambda M$ has a continuous Df-invariant splitting $E \oplus F$ and there exist constants $C > 0$ and $0 < \lambda < 1$ such that

$$\left\|Df^n|_{E(x)}\right\| \cdot \left\|Df^{-n}|_{F(f^n(x))}\right\| \leq C\lambda^n$$

for all $x \in \Lambda$ and $n \geq 0$.

If Λ admits a dominated splitting $T_\Lambda M = E \oplus F$ such that $\dim E(x)$ is constant for $x \in \Lambda$, then there exists a C^1 neighborhood $\mathcal{U}(f)$ and a compact neighborhood W of Λ such that for any $g \in \mathcal{U}(f)$, $\Lambda_g(W)$ admits a dominated splitting (for g)

$$T_{\Lambda_g(W)}M = \hat{E}(g) \oplus \hat{F}(g)$$

with $\dim \hat{E}(g) = \dim E$ (see [11, B.1]).

Let $0 \le j \le \dim M$. Denote by $P_j(f|_\Lambda)$ the set of periodic points $q \in \Lambda \cap \mathrm{Per}(f)$ with $\dim E^s(q) = j$.

Note that both $P_0(f|_\Lambda)$ and $P_{\dim M}(f|_\Lambda)$ are (single) periodic orbits if $f|_\Lambda$ is transitive.

In the next two propositions, assume that f is C^1-stably shadowing in U and let $\mathcal{U}(f)$ be given by Lemma 4.2.2.

Then there is a C^1 neighborhood $\mathcal{V}(f)$ of f such that the family of periodic sequences of linear isomorphisms of tangent spaces of M generated by the differentials Dg, where $g \in \mathcal{V}(f)$, along hyperbolic periodic points $q \in \Lambda_g(U) \cap \mathrm{Per}(g)$ is uniformly hyperbolic (see the paragraph located before Proposition 3.2.1). Note that here we consider periodic orbits of g contained in U.

Since in the proof of [42, Proposition II.1], perturbations are done in a small neighborhood of Λ, we can readily obtain the following proposition which is a semi-local variant of Proposition 3.2.1.

Proposition 4.2.1 *Under the above notation and assumptions, there are constants $C > 0$, $m > 0$, and $0 < \lambda < 1$ such that:*

(a) if $g \in \mathcal{V}(f)$, $q \in \Lambda_g(U) \cap P(g)$, and $\pi(q) \ge m$, then

$$\prod_{i=0}^{k-1} \|Dg^m|_{E^s(g^{im}(q))(g)}\| \le C\lambda^k \text{ and } \prod_{i=0}^{k-1} \|Dg^{-m}|_{E^u(g^{-im}(q))(g)}\| \le C\lambda^k,$$

where $k = [\pi(q)/m]$;

(b) if $g \in \mathcal{V}(f)$ and $0 < j < \dim M$, then $\overline{P_j(g|_{\Lambda_g(U)})}$ admits a dominated splitting $T_{\overline{P_j(g|_{\Lambda_g(U)})}}M = E(g) \oplus F(g)$ with $\dim E(g) = j$, i.e.,

$$\|Dg^m|_{E(x)(g)}\| \cdot \|Dg^{-m}|_{F(g^m(x))(g)}\| \le \lambda$$

for all $x \in \overline{P_j(g|_{\Lambda_g(U)})}$ (note that $E(x)(g) = E^s(x)(g)$ and $F(x)(g) = E^u(x)(g)$ if $x \in P_j(g|_{\Lambda_g(U)})$).

We construct the dominated splitting on the chain transitive set Λ by employing a stronger variant of Pugh's Closing Lemma proved in [42] under the condition that $f|_\Lambda$ is C^1-stably shadowing.

The above proposition will play an essential part in that proof.

Since it is still unknown at this stage whether there is a periodic point in Λ, we cannot apply the same reasoning as in Chap. 3 to prove the hyperbolicity of Λ (even if there exists a dominated splitting on Λ).

It is easy to see that the above proposition can be restated in the following form, which will be used in the proof of Theorem 4.2.1.

Proposition 4.2.2 *Under the notation and assumptions of Proposition 4.2.1, there are constants $m > 0$, $0 < \lambda < 1$, and $L > 0$ such that:*

(a) if $g \in \mathscr{V}(f)$, $q \in \Lambda_g(U) \cap P(g)$, and $\pi(q) \geq L$, then

$$\prod_{i=0}^{\pi(q)-1} \|Dg^m|_{E^s(g^{im}(q))(g)}\| < \lambda^{\pi(q)} \text{ and } \prod_{i=0}^{\pi(q)-1} \|Dg^{-m}|_{E^u(g^{-im}(q))(g)}\| < \lambda^{\pi(q)};$$

(b) if $g \in \mathscr{V}(f)$ and $0 < j < \dim M$, then $\overline{P_j(g|_{\Lambda_g(U)})}$ admits a dominated splitting $T_{\overline{P_j(g|_{\Lambda_g(U)})}}M = E(g) \oplus F(g)$ with $\dim E(g) = j$ such that

$$\|Dg^m|_{E(x)(g)}\| \cdot \|Dg^{-m}|_{F(g^m(x))(g)}\| < \lambda^2$$

for any $x \in \overline{P_j(g|_{\Lambda_g(U)})}$ (note that $E(x)(g) = E^s(x)(g)$ and $F(x)(g) = E^u(x)(g)$ if $x \in P_j(g|_{\Lambda_g(U)})$).

4.2.2 Construction of the Dominated Splitting and Its Extension

Let $f|_\Lambda$ be chain transitive and C^1-stably shadowing. In this subsection, we apply Pugh's Closing Lemma to construct a dominated splitting on Λ and then extend it continuously to a neighborhood of Λ.

First of all, let us state some lemmas which we need.

Denote by $B_\varepsilon(f,x)$ the ε-tubular neighborhood of the f-orbit of x:

$$B_\varepsilon(f,x) = \{y \in M : \operatorname{dist}(f^n(x), y) < \varepsilon \text{ for some } n \in \mathbb{Z}\}.$$

The next lemma is a stronger variant of Pugh's Closing Lemma proved by Mañé (see [42, Lemma I.2]).

Lemma 4.2.3 *Let $f \in \operatorname{Diff}^1(M)$, $x \in M$, $\varepsilon > 0$, and $\mathscr{U}(f)$ be given.*

Then there are $r > 0$ and $\rho > 1$ such that if $y \in N(\bar{r}, x)$ with $0 < \bar{r} \leq r$ and $f^n(y) \in N(\bar{r}, x)$ for some $n > 0$, then there exist $0 \leq n_1 < n_2 \leq n$ and $g \in \mathscr{U}(f)$ such that

- *$f^{n_1}(y) \in N(\rho\bar{r}, x)$ and $f^{n_2}(y) \in N(\rho\bar{r}, x)$;*
- *$g^{n_2-n_1}(f^{n_2}(y)) = f^{n_2}(y) \in Per(g)$;*

– $g = f$ on $M \setminus B_\varepsilon(f, x)$;
  $\text{dist}\left(g^i(f^{n_2}(y)), f^i(f^{n_1}(y))\right) \leq c$ for all $0 \leq i \leq n_2 - n_1$.

In the following two lemmas, we denote by $\mathcal{U}_\varepsilon(f)$ the ε-ball centered at f in $\text{Diff}^1(M)$ with respect to the C^1 metric on $\text{Diff}^1(M)$ and by dist_H the Hausdorff metric on the space of nonempty closed subsets of M.

Lemma 4.2.4 *Assume that $f|_\Lambda$ is transitive. Then for any $n > 0$ there are $g_n \in \mathcal{U}_{1/n}(f)$ and $p_n \in Per(g_n)$ with $\pi(p_n) \geq n$ such that*

$$\text{dist}_H(O(p_n, g_n), \Lambda) < 1/n.$$

Proof Recall that Λ is a nontrivial set, i.e., Λ is not a periodic orbit. Since $f|_\Lambda$ is transitive, there is a point $x \in \Lambda$ such that $\omega_f(x) = \Lambda$ (of course, $x \notin \text{Per}(f)$).

Thus, for any $n > 0$ there is $m_n \geq n$ such that for any $z \in \Lambda$, there is a number $0 \leq j \leq m_n$ for which $\text{dist}(z, f^j(x)) < 1/3n$.

Choose a small enough $\varepsilon_n > 0$ such that the inequality $\text{dist}(x, y) < \varepsilon_n$ (where $y \in \Lambda$) implies that $\text{dist}\left(f^i(x), f^i(y)\right) < 1/3n$ for all $1 \leq i \leq m_n$.

Let $r_n > 0$ and $\rho_n > 1$ be the numbers given by Lemma 4.2.3 for the above x, $1/3n$, and $\mathcal{U}_{1/3n}(f)$.

Take $0 < \bar{r}_n \leq r_n$ such that $\rho_n \bar{r}_n < \varepsilon_n$.

Then, since $\omega_f(x) = \Lambda$, there exist $w \in O(x, f)$ and $l_n > 0$ such that

$$w \in N(\bar{r}_n, x) \text{ and } f^{l_n}(w) \in N(\bar{r}_n, x).$$

By Lemma 4.2.3, there exist $0 \leq l_n^1 < l_n^2 \leq l_n$ and $g_n \in \mathcal{U}_{1/3n}(f)$ such that

– $f^{l_n^1}(w) \in N(\varepsilon_n, x)$ and $f^{l_n^2}(w) \in N(\varepsilon_n, x)$;
– $g_n^{l_n^2 - l_n^1}(f^{l_n^2}(w)) = f^{l_n^2}(w) \in \text{Per}(g_n)$;
– $g_n = f$ on $M \setminus B_{1/3n}(f, f^{l_n^1}(w))$;
– $\text{dist}(f^i(f^{l_n^1}(w)), g_n^i(f^{l_n^2}(w))) \leq 1/3n$ for all $0 \leq i \leq l_n^2 - l_n^1$.

Since $x \notin \text{Per}(f)$, we may assume that $l_n^2 - l_n^1 \geq m_n$.

Therefore, if we put $v_n = f^{l_n^1}(w) \in \Lambda$ and $p_n = f^{l_n^2}(w) \in \text{Per}(g_n)$, then

$$\text{dist}(z, g_n^j(p_n)) \leq \text{dist}(z, f^j(x)) + \text{dist}(f^j(x), f^j(v_n)) + \text{dist}(f^j(v_n), g_n^j(p_n)) < \frac{1}{n}$$

for the above $z \in \Lambda$. Hence, $O(p_n, g_n) \cap N(1/n, z) \neq \emptyset$.

Obviously,

$$O(p_n, g_n) \subset N(1/3n, \Lambda) \subset N(1/n, \Lambda)$$

since $f^i(v_n) \in \Lambda$ for $0 \leq i \leq \pi(p_n)$, where $\pi(p_n) = l_n^2 - l_n^1$.

It is easy to see that p_n can be chosen so that $\pi(p_n)$ is arbitrarily large as $n \to \infty$; thus, we may assume that $\pi(p_n) \geq n$. □

In what follows, we assume in this subsection that $f|_{\Lambda}$ is chain transitive and $f|_{\Lambda}$ is C^1-stably shadowing in U. Let us construct a dominated splitting on Λ by using Proposition 4.2.2.

Lemma 4.2.5 (Existence of a Dominated Splitting) *Under the above notation and assumptions, there exist constants $m > 0$ and $0 < \lambda < 1$ and a Df-invariant splitting $T_{\Lambda}M = E \oplus F$ such that*

$$\left\|Df^m|_{E(x)}\right\| \cdot \left\|Df^{-m}|_{F(f^m(x))}\right\| \le \lambda$$

for any $x \in \Lambda$.

Proof Since $f|_{\Lambda}$ is C^1-stably shadowing in U and transitive, item (*a*) of Proposition 4.2.2 and Lemma 4.2.4 imply that there are sequences of diffeomorphisms g_n and hyperbolic periodic points $p_n \in \mathrm{Per}(g_n)$ such that $g_n \to f$ with respect to the C^1 topology as $n \to \infty$ and $O(p_n, g_n) \to \Lambda$ as $n \to \infty$ with respect to the Hausdorff metric.

We may assume that the indices of $\{p_n\}$ are constant, say, $0 \le j_0 \le \dim M$.

Observe that by item (*b*) of Proposition 4.2.2, there are constants $m > 0$ and $0 < \lambda < 1$ such that if n is sufficiently large, then

$$\left\|Dg_n{}^m|_{E^s(q)(g_n)}\right\| \cdot \left\|Dg_n{}^{-m}|_{E^u(g_n^m(q))(g_n)}\right\| \le \lambda$$

for any $q \in O(p_n, g_n)$.

Let $\Lambda_0 \subset \Lambda$ be a subset such that for any $x \in \Lambda$, the f-orbit of x, $O(x,f)$, intersects Λ_0 at exactly one point.

For any $x \in \Lambda_0$ we can choose a sequence $q_n \in O(p_n, g_n)$ such that $q_n \to x$ as $n \to \infty$.

Set

$$E(x) = \lim_{n\to\infty} E^s(q_n)(g_n) \;\; \text{and} \;\; F(x) = \lim_{n\to\infty} E^u(q_n)(g_n)$$

by taking a subsequence of $\{q_n\}$, if necessary.

For any $x \in \Lambda \setminus \Lambda_0$ such that $f^i(x) \in \Lambda_0$ for some $i \in \mathbb{Z}$, we put

$$E(x) = Df^{-i}\left(f^i(x)\right) E\left(f^i(x)\right) \;\; \text{and} \;\; F(x) = Df^{-i}\left(f^i(x)\right) F\left(f^i(x)\right).$$

Then, following the reasoning of the proof of [41, Proposition 1.3], we can show that the subbundles E and F on Λ are well-defined (i.e., they do not depend on the choices of $\{g_n\}$ and $\{q_n\}$) and that $E(x) \cap F(x) = \{0\}$ for any $x \in \Lambda$.

Furthermore, it follows from our construction that

$$\|Df^m|_{E(x)}\| \cdot \|Df^{-m}|_{F(f^m(x))}\| \le \lambda$$

for any $x \in \Lambda$. Thus, $T_{\Lambda}M = E \oplus F$ is a dominated splitting for f with $\dim E = j_0$.

Note that j_0 is neither 0 nor $\dim M$ since $f|_\Lambda$ is transitive.

Indeed, if $j_0 = \dim M$, then $f|_\Lambda$ is a contraction, so that there are constants $\delta_0 > 0$ and $0 < \mu < 1$ such that if $x, y \in \Lambda$ and $\text{dist}(x, y) < d_0$, then $\text{dist}(f(x), f(y)) \leq \mu \text{dist}(x, y)$. Since $f|_\Lambda$ is transitive, we can find $x \in \Lambda$ and $l > 0$ such that $f^l(N(d_0, x)) \subset N(d_0, x)$.

Thus, there exists a sink $p \in B_{d_0}(x) \cap \text{Per}(f) \cap \Lambda$ (recall that Λ is locally maximal in U).

This is a contradiction since $f|_\Lambda$ is transitive; thus, $j_0 \neq \dim M$ (and a similar reasoning for $f^{-1}|_\Lambda$ shows that $j_0 \neq 0$). □

Hence, by Lemma 4.2.5, Λ admits a dominated splitting with respect to f^m. In the rest of this section, we prepare one technical lemma dealing with extension of a dominated splitting on small neighborhoods of both Λ and f in M and $\text{Diff}^1(M)$, respectively. To simplify notation, denote f^m by f.

It is known that, if a neighborhood U of Λ is small enough, then there exists a constant $\hat{\lambda} > 0$ with $\lambda < \hat{\lambda} < 1$ and a continuous splitting $T_U M = \hat{E} \oplus \hat{F}$ with $\dim \hat{E} = \dim E = j_0$ such that

- $\hat{E}|_\Lambda = E$ and $\hat{F}|_\Lambda = F$;
- $Df(x)\hat{E}(x)) = \hat{E}(f(x))$ if $x \in U \cap f^{-1}(U)$;
- $Df^{-1}(x)\hat{F}(x) = \hat{F}(f^{-1}(x))$ if $x \in U \cap f(U)$;
- $\left\|Df^k|_{\hat{E}(x)}\right\| \cdot \left\|Df^{-k}|_{\hat{F}(f^k(x))}\right\| < \hat{\lambda}^k$ if $x \in \bigcap_{i=-k}^{k} f^i(U)$ for $k \geq 0$.

Using this continuous splitting, we can prove the following lemma applying the reasoning developed in [27]. This lemma will be used in the proof of Theorem 4.2.1 (more precisely, in the proof of Proposition 4.2.4).

Lemma 4.2.6 (Extension of the Dominated Splitting) *Under the above notation and assumptions, for any $\varepsilon > 0$ there is $d > 0$ such that $N(d, \Lambda) \subset U$ and for any $g \in \mathscr{U}_d(f)$ there is a Dg-invariant continuous splitting*

$$T_{\Lambda_g(N(d,\Lambda))}M = \hat{E}(g) \oplus \hat{F}(g)$$

with $\dim \hat{E}(g) = j_0$ and the following properties:

- $\left\|Dg^k|_{\hat{E}(x)(g)}\right\| \cdot \left\|Dg^{-k}|_{\hat{F}(g^k(x))(g)}\right\| < \hat{\lambda}^k$ *for any $x \in \Lambda_g(N(d, \Lambda))$ and $k \geq 0$;*
- *if $x \in \Lambda_g(N(d, \Lambda))$, $y \in \Lambda$, and $\text{dist}(x, y) < d$, then*

$$\left|\log \|Dg|_{\hat{E}(x)(g)}\| - \log \|Df|_{E(y)}\|\right| < \varepsilon$$

and

$$\left|\log \|Dg^{-1}|_{\hat{F}(x)(g)}\| - \log \|Df^{-1}|_{F(y)}\|\right| < \varepsilon.$$

Here $\Lambda_g(N(d, \Lambda)) = \bigcap_{n\in\mathbb{Z}} g^n(N(d, \Lambda))$.

Finally, let us remark that it is easy to prove that if, under the assumptions of the above lemma, a point $p \in \Lambda_g(N(d, \Lambda)) \cap \mathrm{Per}(g)$ is hyperbolic with $\dim E^s(p)(g) = j_0$, then $E^s_p(g) = \hat{E}(p)(g)$ and $E^u(p)(g) = \hat{F}(p)(g)$ by the "uniqueness" of the dominated splitting (see [11, B.1]).

4.2.3 *Proof of Theorem 4.2.1*

In this section, we prove Theorem 4.2.1 using the idea of the proof of Theorem 3.2.1. The following proposition was already proved in Chap. 3 (see Proposition 3.2.3).

Proposition 4.2.3 *Let $f|_\Lambda$ and $0 < \lambda < 1$ be given; assume that there is a continuous Df-invariant splitting $T_\Lambda M = E \oplus F$ such that*

$$\|Df|_{E(x)}\| \cdot \|Df^{-1}|_{F(f(x))}\| < \lambda^2$$

for any $x \in \Lambda$. Assume that there is a point $y \in \Lambda$ such that

$$\log \lambda < \log \lambda_1 = \limsup_{n\to\infty} \frac{1}{n} \sum_{i=0}^{n-1} \log \|Df|_{E(f^i(y))}\| < 0$$

and

$$\liminf_{n\to\infty} \frac{1}{n} \sum_{i=0}^{n-1} \log \|Df|_{E(f^i(y))}\| < \log \lambda_1.$$

Then for any λ_2 and λ_3 such that

$$\lambda < \lambda_2 < \lambda_1 < \lambda_3 < 1$$

and for any neighborhood U of Λ there is a hyperbolic periodic point q of index $\dim E$ *such that $O(q,f) \subset U$*

$$\prod_{i=0}^{k-1} \left\|Df|_{E^s(f^i(q))}\right\| \le \lambda_3^k, \text{ and } \prod_{i=k-1}^{\pi(q)-1} \left\|Df|_{E^s(f^i(q))}\right\| > \lambda_2^{\pi(q)-k+1}$$

for all $k = 1, 2, \dots, \pi(q)$. Furthermore, q can be chosen so that $\pi(q)$ is arbitrarily large.

The main auxiliary statement which we prove in this section is the following proposition. The proof slightly modifies the reasoning used in the proof of the main result of [109] and Chap. 3 (the main modification in our proof is the construction of an extra pseudotrajectory reflecting the assumed nonhyperbolicity, see the paragraph located before Step I of the proof).

Finally, the "only if" part of Theorem 4.2.1 will be obtained by showing that f has properties (P.1)–(P.5) of Proposition 4.2.4 if $f|_\Lambda$ is chain transitive and C^1-stably shadowing.

Proposition 4.2.4 *Let $f|_\Lambda$ be locally maximal in U and let $0 < \lambda < 1$ and $L > 1$ be given. Assume that $f|_\Lambda$ has the following properties* (P.1)–(P.5)*:*

(P.1) *There is a Df-invariant splitting $T_\Lambda M = E \oplus F$ such that if $x \in \Lambda$, then*

$$\|Df|_{E(x)}\| \cdot \|Df^{-1}|_{F(f(x))}\| < \lambda^2.$$

(P.2) *There is a neighborhood $\mathcal{U}(f)$ such that if $g \in \mathcal{U}(f)$, $q \in \Lambda_g(U) \cap Per(g)$, and $\pi(q) \geq L$, then*

$$\prod_{i=0}^{\pi(q)-1} \|Dg|_{E^s(g^i(q))(g)}\| < \lambda^{\pi(q)} \text{ and } \prod_{i=0}^{\pi(q)-1} \|Dg^{-1}|_{E^u(g^{-i}(q))(g)}\| < \lambda^{\pi(q)}.$$

(P.3) *For any $\varepsilon > 0$ with $N(\varepsilon, \Lambda) \subset U$ and $\mathcal{U}_\varepsilon(f) \subset \mathcal{U}(f)$ there exist $g \in \mathcal{U}_\varepsilon(f)$ and $p \in Per(g)$ such that $\mathrm{dist}_H(O(p, g), \Lambda) < \varepsilon$. Furthermore, p can be chosen so that $\pi(p)$ is arbitrarily large.*

(P.4) *For any $\varepsilon > 0$ there is $d > 0$ with $N(d, \Lambda) \subset U$ and $\mathcal{U}_d(f) \subset \mathcal{U}(f)$ such that if $g \in \mathcal{U}_d(f)$, $p \in \Lambda_g(N(d, \Lambda)) \cap Per(g)$, $y \in \Lambda$, and $\mathrm{dist}(p, y) < d$, then*

$$\left| \log \|Dg|_{E^s(p)(g)}\| - \log \|Df|_{E(y)}\| \right| < \varepsilon$$

and

$$\left| \log \|Dg^{-1}|_{E^u(p)(g)}\| - \log \|Df^{-1}|_{F(y)}\| \right| < \varepsilon.$$

(P.5) *$f|_\Lambda$ is shadowing.*

Then Λ is hyperbolic.

Proof Let $f|_\Lambda$ be locally maximal in U and let $0 < \lambda < 1$ and $L > 0$ be given. Assume that $f|_\Lambda$ has properties (P.1)–(P.5) and let $T_\Lambda M = E \oplus F$ be a Df-invariant splitting as in (P.1) (recall that a dominated splitting is continuous).

Assuming that E is not contracting, we show first that for any $\lambda < \eta < \eta' < 1$ there is a point $z \in \Lambda$ such that

$$\liminf_{n\to\infty} \frac{1}{n} \sum_{j=0}^{n-1} \log \|Df|_{E(f^j(z))}\| < \log \eta < \limsup_{n\to\infty} \frac{1}{n} \sum_{j=0}^{n-1} \log \|Df|_{E(f^j(z))}\| < \log \eta'.$$

After that, we get a contradiction applying Proposition 4.2.3.

It is known that if there exists $N > 0$ such that for any $x \in \Lambda$ there is a number $0 \leq n(x) \leq N$ for which $\|Df^{n(x)}|_{E(x)}\| < 1$, then E is contracting.

Since E is not contracting, it is easy to see that there is a point $y \in \Lambda$ such that

$$\prod_{j=0}^{n-1} \left\| Df|_{E(f^j(y))} \right\| \geq 1 \text{ for all } n \geq 1.$$

Let $\mathscr{U}(f)$ be as in property (P.2); choose $\varepsilon > 0$ small enough so that $N(\varepsilon, \Lambda) \subset U$, $\mathscr{U}_\varepsilon(f) \subset \mathscr{U}(f)$, and the following conditions are satisfied:

(*i*) if $x, y \in \Lambda$ and $\text{dist}(x, y) < \epsilon$, then

$$\left| \log \left\| Df|_{E(x)} \right\| - \log \left\| Df|_{E(y)} \right\| \right| < \min \left\{ \frac{1}{2}(\log \eta' - \log \eta), \frac{1}{4}(\log \eta - \log \lambda) \right\};$$

(*ii*) if $g \in \mathscr{U}_\varepsilon(f)$, $q \in \Lambda_g(N(\varepsilon, \Lambda)) \cap \text{Per}(g)$, $y \in \Lambda$, and $\text{dist}(q, y) < \epsilon$, then

$$\left| \log \left\| Dg|_{E^s(q)(g)} \right\| - \log \left\| Df|_{E(y)} \right\| \right| < \frac{1}{4}(\log \eta - \log \lambda).$$

Note that the possibility of finding ε for which item (*i*) is satisfied follows from the continuity of E; for item (*ii*), it follows from property (P.4).

Since $f|_\Lambda$ is shadowing, there is $0 < d \leq \varepsilon$ such that any d-pseudotrajectory of f in Λ can be ε-shadowed by a trajectory of f.

Since $\text{dist}(f(x), f(y)) \leq e^K \text{dist}(x, y)$ for any $x, y \in M$, where

$$K = \max \left\{ \left| \log \|Df(x)\| \right| : x \in M \right\},$$

it is not hard to show that there exists a number $0 < \nu \leq d/2$ such that if $x, y \in M$, $g \in \mathscr{U}_\nu(f)$, and $\text{dist}(x, y) < \nu$, then

$$\text{dist}(f(x), g(y)) < \frac{d}{2}.$$

By property (P.3), there exists a diffeomorphism $g \in \mathscr{U}_\nu(f)$ and a hyperbolic periodic point p of g with $\pi(q) \geq L$ such that its g-orbit, $O(p, g)$, forms a ν-net of Λ, i.e., for any point $w \in \Lambda$ there is a point $q \in O(p, g)$ such that $\text{dist}(w, q) < \nu$, and, conversely, for any $q \in O(p, g)$, there is $w \in \Lambda$ such that $\text{dist}(w, q) < \nu$.

At first, let us construct a periodic d-pseudotrajectory of f in Λ (with period $\pi(p)$) that approximates the above periodic orbit $O(p, g)$ of g within ν with respect to dist_H.

Take points $q_j \in \Lambda$ such that $\text{dist}(g^j(p), q_j) < \nu$ for $j = 0, 1, \ldots, \pi(p) - 1$.

Then

$$\mathrm{dist}(f(q_j), q_{j+1}) \le \mathrm{dist}(f(q_j), g(g^j(p))) + \mathrm{dist}(g^{j+1}(p), q_{j+1}) < \frac{d}{2} + \nu \le d,$$

and

$$(iii)\quad \frac{1}{\pi(p)} \sum_{j=0}^{\pi(p)-1} \log \|Df|_{E(q_j)}\| < \frac{1}{2}(\log\lambda + \log\eta)$$

by the choice of ν.

Thus, the sequence of points $\{q_j\}_{j=0}^{\pi(p)-1} \subset \Lambda$ is a periodic d-pseudotrajectory of f; in what follows, we denote it by $\mathcal{PO}_f(p)$.

Remark that in the proof of Theorem 3.2.1 (Proposition 3.2.4), the above orbit was an exact periodic orbit of f; however, in our case, it is a periodic pseudotrajectory; this is the main difference between the original proof in [109] and our proof.

Observe that the inequality in (*iii*) follows from (P.2) and (*ii*) since

$$\frac{1}{\pi(p)} \sum_{j=0}^{\pi(p)-1} \log \|Dg|_{E^s(g^j(p))(g)}\| < \frac{1}{4}(3\log\lambda + \log\eta).$$

We will construct a d-pseudotrajectory

$$\{x_i\}_{i\in\mathbb{Z}} \subset \Lambda$$

of f composed of points of $O(y,f)$ and $\mathcal{PO}_f(p)$ by mimicking the procedure displayed in the proof of the main result of Proposition 3.2.4 (see also [109]). □

Step I Since $y \in \Lambda$, there is a point $q_{j_1} \in \mathcal{PO}_f(p)$ such that $\mathrm{dist}(y, q_{j_1}) < \nu < d$. Set

$$x_{-1} = q_{j_1-1},\ x_{-2} = q_{j_1-2},\ \ldots,\ x_{-\pi(p)+1} = q_{j_1-\pi(p)+1},$$

and

$$x_{-\pi(p)} = q_{j_1},\ x_{-\pi(p)-1} = q_{j_1-1},\ x_{-\pi(p)-2} = q_{j_1-2},\ \ldots.$$

Then $\mathrm{dist}(f(x_{-i}), x_{-i+1}) < d$ for $i \ge 1$, so that the negative part $\{x_i\}_{i=-\infty}^{-1}$ of $\{x_i\}_{i\in\mathbb{Z}}$ is constructed.

Step II Let $n_1 = 1$. Then

$$\frac{1}{n_1\pi(p)}\left(n_1 \sum_{j=0}^{\pi(p)-1} \log \left\|Df|_{E(q_{j_1+j})}\right\|\right) < \frac{1}{2}(\log\lambda + \log\eta).$$

Obviously, this inequality follows from (*iii*).

Let $i_1 = n_1\pi(p)$, put $x_j = q_{j_1+j}$ for $j = 0, 1, \ldots, i_1 - 1 = \pi(p) - 1$, and put $x_{i_1} = y$. Then $\mathrm{dist}(f(x_j), x_{j+1}) < d$ for $j = 0, 1, \ldots, i_1 - 1$, and

$$\frac{1}{i_1}\sum_{j=0}^{i_1-1}\log\left\|Df|_{E(x_j)}\right\| < \frac{1}{2}(\log\lambda + \log\eta).$$

Put

$$a_j = \log\|Df|_{E(x_j)}\|$$

for $j = 0, 1, \ldots, i_1 - 1$ and choose a number l_1 having the following properties:

$$\frac{1}{i_1 + l_1}\left(\sum_{j=0}^{i_1-1} a_j + \sum_{j=0}^{l_1-1}\log\left\|Df|_{E(f^j(y))}\right\|\right) \geq \frac{1}{2}(\log\eta + \log\eta')$$

and

$$\frac{1}{i_1 + l}\left(\sum_{j=0}^{i_1-1} a_j + \sum_{j=0}^{l-1}\log\left\|Df|_{E(f^j(y))}\right\|\right) < \frac{1}{2}(\log\eta + \log\eta')$$

for any $l < l_1$.

The existence of l_1 follows from our choice of y (recall that

$$\prod_{j=0}^{n-1}\left\|Df|_{E(f^j(y))}\right\| \geq 1 \text{ for all } n \geq 1).$$

Set $j_1 = i_1 + l_1$, let $x_{i_1+1} = f(y), x_{i_1+2} = f^2(y), \ldots, x_{j_1-1} = f^{l_1-1}(y) \in O(y, f)$, and put

$$a_{i_1+j} = \log\left\|Df|_{E(x_{i_1+j})}\right\|$$

for $j = 0, 1, \ldots, l_1 - 1$.

Step III Let $i_{k-1}, j_{k-1}, \{x_i\}_{i=0}^{j_{k-1}-1}$, and $\{a_i\}_{i=0}^{j_{k-1}-1}$ have been constructed in the former steps. Similarly to the choice of q_{j_1} and n_1, we can choose $q_{j_k} \in \mathscr{P}\mathscr{O}_f(p)$ such that

$$\mathrm{dist}(f(x_{j_{k-1}}), q_{j_k}) < \nu < d$$

and a positive number n_k such that

$$\frac{1}{i_k}\left(\sum_{j=0}^{j_{k-1}-1} a_j + n_k \sum_{j=0}^{\pi(p)-1} \log \left\| Df|_{E(q_{j_k+j})} \right\| \right) < \frac{1}{2}(\log\lambda + \log\eta),$$

where $i_k = j_{k-1} + n_k\pi(p)$ (the existence of n_k is ensured by (*iii*)).

Let

$$x_{j_{k-1}+1} = q_{j_k+1},\ x_{j_{k-1}+2} = q_{j_k+2},\ \ldots,\ x_{j_{k-1}+\pi(p)} = q_{j_k},$$
$$x_{j_{k-1}+\pi(p)+1} = q_{j_k+1},\ x_{j_{k-1}+\pi(p)+2} = q_{j_k+2},\ \ldots,$$

and $x_{i_k} = f(x_{j_{k-1}-1}) \in O(y,f)$.

Obviously,

$$\mathrm{dist}\left(f(x_{j_{k-1}+j}), x_{j_{k-1}+j+1}\right) < d$$

for $j = 0, 1, \ldots, n_k\pi(p) - 1$.

Put

$$a_{j_{k-1}+j} = \log \left\| Df|_{E(x_{j_{k-1}+j})} \right\|$$

for $j = 0, 1, \ldots, n_k\pi(p) - 1$ and choose l_k such that

$$\frac{1}{i_k + l_k}\left(\sum_{j=0}^{i_k-1} a_j + \sum_{j=0}^{l_k-1} \log \left\| Df|_{E(f^j(x_{i_k}))} \right\| \right) \geq \frac{1}{2}(\log\eta + \log\eta')$$

and

$$\frac{1}{i_k + l}\left(\sum_{j=0}^{i_k-1} a_j + \sum_{j=0}^{l} \log \left\| Df|_{E(f^j(x_{i_k}))} \right\| \right) < \frac{1}{2}(\log\eta + \log\eta')$$

for any $l < l_k$.

The existence of l_k is ensured by the fact that $x_{i_k} \in O(y,f)$ (recall the choice of y).

Let $j_k = i_k + l_k$ and let $x_{i_k+1} = f(x_{i_k}), x_{i_k+2} = f^2(x_{i_k}), \ldots, x_{j_k-1} = f^{l_k-1}(x_{i_k})$. Finally, we put

$$a_{j_{k-1}+j} = \log \left\| Df|_{E(f^j(x_{i_k}))} \right\|$$

for $j = 0, 1, \ldots, l_k - 1$.

This completes the construction of $\{x_i\}_{i\in\mathbb{Z}} \subset \Lambda$ of f. Roughly speaking, the d-pseudotrajectory $\{x_i\}_{i\in\mathbb{Z}}$ looks as follows:

$$\{\ldots,\ \mathscr{P}\mathscr{O}_f(p),\ \mathscr{P}\mathscr{O}_f(p),\ y,\ f(y),\ f^2(y),\ \ldots,\ f^{l_1}(y),\ \mathscr{P}\mathscr{O}_f(p),$$

$$\ldots,\ \mathscr{P}\mathscr{O}_f(p),\ f^{l_1+1}(y),\ \ldots,\ f^{l_1+l_2}(y),\ \mathscr{P}\mathscr{O}_f(p),\ \ldots\}.$$

Recall that $K = \max\{|\log\|Df(x)\|| : x \in M\}$. It is easy to see that

$$\frac{1}{i_k}\sum_{j=0}^{i_{k-1}-1} a_j < \frac{1}{2}(\log\lambda + \log\eta) \text{ and } \frac{1}{j_k}\sum_{j=0}^{j_k-1} a_j \geq \frac{1}{2}(\log\eta + \log\eta')$$

for every $k = 1, 2, \ldots$, and

$$\frac{1}{n}\sum_{j=0}^{n-1} a_j < \frac{1}{n}\left(\frac{1}{2}(\log\eta + \log\eta')\,(n - \pi(p)) + K\pi(p)\right)$$

for every $n \geq \pi(p)$.

Hence,

$$\limsup_{n\to\infty}\frac{1}{n}\sum_{j=0}^{n-1} a_j = \frac{1}{2}(\log\eta + \log\eta')$$

and

$$\liminf_{n\to\infty}\frac{1}{n}\sum_{j=0}^{n-1} a_j \leq \frac{1}{2}(\log\lambda + \log\eta).$$

Let $z \in M$ be a point such that its f-orbit ε-shadows the pseudotrajectory $\{x_i\}_{i\in\mathbb{Z}}$ (see (P.5)).

Observe that $O(z,f) \subset U$, so that $z \in \Lambda$ by the local maximality.

Thus, the choice of ε (see (i)) implies that

$$\liminf_{n\to\infty}\frac{1}{n}\sum_{j=0}^{n-1}\log\left\|Df|_{E(f^j(z))}\right\| < \log\eta < \limsup_{n\to\infty}\frac{1}{n}\sum_{j=0}^{n-1}\log\left\|Df|_{E(f^j(z))}\right\| < \log\eta'.$$

Thus, by Proposition 4.2.3, there is a hyperbolic periodic point q of index $\dim E$ such that $O(q,f) \subset U$ and the derivatives along the orbit $O(q,f)$ satisfy the following inequalities:

$$\prod_{i=0}^{k-1} \|Df|_{E^s(f^i(q))}\| \le \eta'^k \text{ and } \prod_{i=k-1}^{\pi(q)-1} \|Df|_{E^s(f^i(q))}\| > \eta^{\pi(q)-k+1}$$

for all $k = 1, 2, \ldots, \pi(q)$.

Furthermore, q can be chosen such that $\pi(q)$ is arbitrarily large, so that we may assume that $\pi(q) \ge L$. This is a contradiction because

$$\prod_{i=0}^{\pi(q)-1} \|Df|_{E^s(f^i(q))}\| < \lambda^{\pi(q)}$$

by (P.2). In the same manner, we can show that F is expanding, and thus, Λ is hyperbolic.

Now we complete the proof of Theorem 4.2.1.

Assume that $f|_\Lambda$ is chain transitive and C^1-stably shadowing in U, and let $m > 0$, $0 < \lambda < 1$, and $L > 0$ be constants given by Proposition 4.2.2.

Then it is not difficult to show that the assumption of Proposition 4.2.3 and properties (P.1)–(P.5) of Proposition 4.2.4 are satisfied for $f|_\Lambda$ from Proposition 4.2.2 and Lemmas 4.2.4–4.2.6.

More precisely, (P.3) follows from Lemma 4.2.4 applied to f.

Properties (P.1), (P.2), and (P.4) follow from Proposition 4.2.2 and Lemmas 4.2.4–4.2.6 applied to f^m. Note that $f^m|_\Lambda$ is shadowing since $f|_\Lambda$ is shadowing, and thus, f^m has property (P.5).

Finally, Λ is hyperbolic for f^m if and only if it is hyperbolic for f. Thus, applying Propositions 4.2.3 and 4.2.4 to f^m, we can show that Λ is hyperbolic. Since $f|_\Lambda$ is transitive by Lemma 4.2.1, Λ is a basic set. The "only if" part of Theorem 4.2.1 is proved. □

4.2.4 *Proof of Corollary 4.2.1*

Proof Recall that, in general, a chain recurrence class $\mathscr{C}_f$ of f does not contain periodic points. In this section, we prove Corollary 4.2.1 by applying Theorem 4.2.1.

We need the following two lemmas.

Lemma 4.2.7 *Let Λ be a hyperbolic basic set of $f \in \mathrm{Diff}^1(M)$. If $p \in \Lambda \cap Per(f)$ is a saddle periodic point, then $\Lambda \subset H_f(O(p,f))$.*

Proof Let $p \in \Lambda \cap \mathrm{Per}(f)$ be a saddle periodic point. Since Λ is hyperbolic, there is $\varepsilon_0 > 0$ such that if $\mathrm{dist}(f^n(x), f^n(p)) \le \varepsilon_0$ for $n \ge 0$, then $x \in W^s_{\varepsilon_0}(p)$ (and a similar property holds for $W^u_{\varepsilon_0}(p)$ with respect to f^{-1}).

Let U be a compact neighborhood of Λ in which Λ is locally maximal. Since $f|_\Lambda$ is shadowing, for any $\varepsilon \in (0, \varepsilon_0)$ there is a number $d = d(\varepsilon) > 0$ given by the shadowing property of $f|_\Lambda$ (recall that, by Lemma 4.2.1, we can find a shadowing point in Λ).

Since $f|_\Lambda$ is transitive by Lemma 4.2.1, for any $x \in \Lambda$ there exists a point $y \in N(d,p)$ and numbers $0 < l_1 < l_2$ such that $f^{l_1}(y) \in N(\varepsilon, x)$ and $f^{l_2}(y) \in N(d,p)$.

Put $y_{-i} = f^{-i}(p)$ for $i \geq 0$, $y_i = f^i(y)$ for $0 \leq i \leq l_2$, and $y_i = f^{i-l_2}(p)$ for $i \geq i_2$. Then it is easy to see that $\{y_i\}_{i\in\mathbb{Z}} \subset \Lambda$ is a d-pseudotrajectory of f.

Thus, there is a point $z \in \Lambda$ (ε-close to x) that ε-shadows the pseudotrajectory. Hence,

$$z \in \big(W^s(O(p,f)) \cap W^u(O(p,f))\big) \cap N(\varepsilon, x) \neq \emptyset.$$

Observe that z is a transverse intersection point since Λ is hyperbolic. Since ε is arbitrary, $x \in H_f(O(p,f))$; thus, $\Lambda \subset H_f(O(p,f))$ as claimed. □

Lemma 4.2.8 *Let $\mathscr{C}_f$ be a chain recurrence class of f. If $\mathscr{C}_f$ is hyperbolic, then it is locally maximal.*

Proof Let $\mathscr{C}_f$ be hyperbolic. We first show that for any $\varepsilon > 0$ there is $d > 0$ such that for any d-pseudotrajectory $\{x_i\}_{i\in\mathbb{Z}}$ of f in $\mathscr{C}_f$ there is a point $y \in \mathscr{C}_f$ such that $\mathrm{dist}(f^i(y), x_i) < \varepsilon$ for $i \in \mathbb{Z}$, i.e., the shadowing point y can be taken from $\mathscr{C}_f$.

To prove this, it is enough to show that $f|_{\mathscr{C}_f}$ has a local product structure.

Since $\mathscr{C}_f$ is hyperbolic, for any $\varepsilon > 0$ there is $d > 0$ such that if $x, y \in \mathscr{C}_f$ and $\mathrm{dist}(x,y) < d$, then $W^s_\varepsilon(x)$ and $W^u_\varepsilon(y)$ have a point of transverse intersection. Fix $x, y \in \mathscr{C}_f$ with $\mathrm{dist}(x,y) < d$ and let

$$z = W^s_\varepsilon(x) \cap W^u_\varepsilon(y) \ \text{ and } \ w = W^u_\varepsilon(x) \cap W^s_\varepsilon(y).$$

We claim that $z, w \in \mathscr{C}_f$.

For any $\eta > 0$ there is $n > 0$ such that

$$\max\{\mathrm{dist}(f^n(x), f^n(z)), \mathrm{dist}(f^{-n}(y), f^{-n}(z))\} < \eta.$$

Since $f^{\pm}(\mathscr{C}_f) = \mathscr{C}_f$, $f^n(x), f^{-n}(y) \in \mathscr{C}_f$.

Thus, $f^n(x) \rightsquigarrow x$ and $x \rightsquigarrow f^{-n}(y)$; i.e., there exist an η-pseudotrajectory $\{x_i\}_{i=0}^{n_\eta}$ with $x_0 = f^n(x)$ and $x_{n_\eta} = x$ and an η-pseudotrajectory $\{y_i\}_{i=0}^{m_\eta}$ with $y_0 = x$ and $y_{m_\eta} = f^{-n}(y)$.

Since η is arbitrary, $z \leftrightsquigarrow x$, so that $z \in \mathscr{C}_f$. A similar reasoning shows that $w \in \mathscr{C}_f$; thus, $f|_{\mathscr{C}_f}$ has a local product structure.

Since $\mathscr{C}_f$ is hyperbolic, there exists $c > 0$ such that if $x \in \mathscr{C}_f$, $y \in M$, and $\mathrm{dist}(f^n(x), f^n(y)) \leq c$ for $n \in \mathbb{Z}$, then $x = y$.

Now fix $0 < \epsilon \leq c/2$ and let $0 < d < \epsilon$ be the corresponding number given by the shadowing property of $f|_{\mathscr{C}_f}$. Furthermore, choose $0 < d' < d/2$ such that if $\mathrm{dist}(x,y) < d'$, then $\mathrm{dist}(f(x), f(y)) < d/2$.

We claim that

$$\mathscr{C}_f = \bigcap_{i\in\mathbb{Z}} f^i(N(d', \mathscr{C}_f)).$$

It is obvious that $\mathscr{C}_f \subset \bigcap_{i\in\mathbb{Z}} f^i(N(d', \mathscr{C}_f))$.

To show the converse, we note that any point $x \in \bigcap_{i\in\mathbb{Z}} f^i(N(d', \mathscr{C}_f))$ is in $\mathscr{C}_f$.

For each $i \in \mathbb{Z}$ take a point $x_i \in \mathscr{C}_f$ such that $\text{dist}(f^i(x), x_i) < d'$. It is easy to see that $\{x_i\}_{i\in\mathbb{Z}}$ is a d-pseudotrajectory of $f|_{\mathscr{C}_f}$ by the choice of d'. By the shadowing property of $f|_{\mathscr{C}_f}$, there is a point $y \in \mathscr{C}_f$ such that $\text{dist}(f^i(y), x_i) < \varepsilon$ for all $i \in \mathbb{Z}$.

Thus,

$$\text{dist}(f^i(x), f^i(y)) \leq \text{dist}(f^i(x), x_i) + \text{dist}(x_i, f^i(y)) < d' + \varepsilon < c$$

for all $i \in \mathbb{Z}$; hence, $x = y \in \mathscr{C}_f$. □

Now we complete the proof of Corollary 4.2.1.

Let $\mathscr{C}_f$ be a chain recurrence class of f and assume that $f|_{\mathscr{C}_f}$ is C^1-stably shadowing in U, i.e., that $f|_{\mathscr{C}_f}$ is locally maximal in U and $f \in \text{Int}^1(\text{SSP}_D(U))$.

By Proposition 1.1.1, $f|_{\mathscr{C}_f}$ is chain transitive, and hence, $\mathscr{C}_f$ is a hyperbolic basic set by Theorem 4.2.1 (recall that periodic points are dense in $\mathscr{C}_f$).

By Lemma 4.2.7, $\mathscr{C}_f \subset H_f(O(p,f))$ for some $p \in \mathscr{C}_f \cap \text{Per}(f)$. Since $f|_{H_f(O(p,f))}$ is transitive, we get the equality $\mathscr{C}_f = H_f(O(p,f))$ because $\mathscr{C}_f$ is a maximal chain transitive set.

The proof of "if" part is as follows. If a chain recurrence class $\mathscr{C}_f$ of f is hyperbolic, then it is locally maximal by Lemma 4.2.8, so that it is a basic set by Lemma 4.2.1 (since $f|_{\mathscr{C}_f}$ is shadowing). Thus, by the local stability of hyperbolic basic sets (see [84]), $f|_{\mathscr{C}_f}$ is C^1-stably shadowing. □

Historical Remarks Theorem 4.2.1 and Corollary 4.2.1 were proved by the second author in [91]. An assertion similar to the corollary was first proved in [35, Theorem I.3] in the case where $\mathscr{C}_f$ is a chain recurrence class containing a periodic point of f. Since, in general, a chain recurrence class does not contain a periodic point, Corollary 4.2.1 generalizes the result of [35].

4.3 Chain Transitive Sets with Shadowing for Generic Diffeomorphisms

Let us recall that a subset $\mathscr{R}$ of $\text{Diff}^1(M)$ is called residual if it is a countable intersection of dense open sets of $\text{Diff}^1(M)$ (see Chap. 1).

In this section, we prove the following C^1-generic result for locally maximal chain transitive sets with shadowing obtained in [34].

Theorem 4.3.1 *There is a residual set $\mathcal{R} \subset \mathrm{Diff}^1(M)$ such that if $f \in \mathcal{R}$ and Λ is a locally maximal chain transitive set of f, then Λ is hyperbolic if and only if $f|_\Lambda$ is shadowing.*

Let Λ be a locally maximal chain transitive set of $f \in \mathcal{R}$. Remark that by the theorem, if $f|_\Lambda$ is shadowing, then Λ is a hyperbolic basic set (see Lemma 4.2.8).

We start with two lemmas which will be used in the proof of Theorem 4.3.1.

First we note that there is a residual set $\mathcal{R}_1 \subset \mathrm{Diff}^1(M)$ such that every $f \in \mathcal{R}_1$ has the following properties.

(1) Every periodic point of f is hyperbolic, and stable and unstable manifolds of periodic points of f are transverse.
(2) A compact f-invariant set Λ is chain transitive if and only if Λ is the limit of a sequence of periodic orbits of f with respect to the Hausdorff distance.

Statement (1) above follows from the Kupka-Smale theorem (see Theorem 1.3.6 (a)). Statement (2) is proved in [16].

Note that if Λ is a locally maximal chain transitive set of $f \in \mathcal{R}_1$, then statement (2) above implies that periodic points are dense in Λ.

Recall that the index of a hyperbolic periodic point $p \in \mathrm{Per}(f)$ is the dimension of the stable manifold of p.

Lemma 4.3.1 *There is a residual set $\mathcal{R}_2 \subset \mathrm{Diff}^1(M)$ such that every $f \in \mathcal{R}_2$ has the following property: If $\Lambda \subset M$ is a closed f-invariant set $\Lambda \subset M$ and there is a sequence of diffeomorphisms f_n converging to f and a sequence of hyperbolic periodic orbits P_n of f_n with index k such that*

$$\lim_{n\to\infty} P_n = \Lambda,$$

then there is a sequence of hyperbolic periodic orbits Q_n of f with index k such that Λ is the Hausdorff limit of Q_n.

Proof Denote by $\mathcal{K}(M)$ be the space of all nonempty compact subsets of M equipped with the Hausdorff metric and take a countable basis $\beta = \{\mathcal{V}_n\}_{n=1}^{\infty}$ of $\mathcal{K}(M)$.

For each pair (n, k) with $n \geq 1$ and $k \geq 0$, we denote by $\mathcal{H}_{n,k}$ the set of f such that f has a C^1 neighborhood $\mathcal{U}(f) \subset \mathrm{Diff}^1(M)$ with the following property: If $g \in \mathcal{U}(f)$, then there exists a hyperbolic periodic orbit $Q \subset \mathcal{V}_n$ of g with index k.

Let $\mathcal{N}_{n,k}$ be the set of f such that f has a neighborhood $\mathcal{U}(f) \subset \mathrm{Diff}^1(M)$ with the following property: Every diffeomorphism $g \in \mathcal{U}(f)$ does not have hyperbolic periodic orbits $Q \subset \mathcal{V}_n$ with index k.

It is clear that the sets $\mathcal{H}_{n,k} \cup \mathcal{N}_{n,k}$ are open in $\mathrm{Diff}^1(M)$.

Let us show that any set $\mathcal{H}_{n,k} \cup \mathcal{N}_{n,k}$ is dense in $\mathrm{Diff}^1(M)$. Take an arbitrary diffeomorphism $f \in \mathrm{Diff}^1(M) \setminus \mathcal{N}_{n,k}$.

Then for any neighborhood $\mathcal{U}(f)$ of f there is a diffeomorphism $g \in \mathcal{U}(f)$ having a hyperbolic periodic orbit $Q \subset \mathcal{V}_n$ with index k. The hyperbolicity of Q

implies that $g \in \mathcal{H}_{n,k}$. This means that $f \in \overline{\mathcal{H}_{n,k}}$; thus,

$$\mathrm{Diff}^1(M) = \overline{\mathcal{H}_{n,k} \cup \mathcal{N}_{n,k}} \subset \overline{\mathcal{H}_{n,k}} \cup \mathcal{N}_{n,k}.$$

Let

$$\mathcal{R}_2 = \bigcap_{n\in\mathbb{Z}^+,\ k=0,\dots,\dim M} \mathcal{H}_{n,k} \cup \mathcal{N}_{n,k}.$$

Then $\mathcal{R}_2$ is a residual subset of $\mathrm{Diff}^1(M)$.

Let $f \in \mathcal{R}_2$ and let Λ be a closed f-invariant subset of M. Assume that there is a sequence of diffeomorphisms f_n converging to f and a sequence of periodic orbits P_n of f_n with index k such that Λ is the Hausdorff limit of P_n.

Fix an arbitrary neighborhood $\mathcal{V}$ of Λ in $\mathcal{K}(M)$ and take $\mathcal{V}_m \in \beta$ such that $\Lambda \subset \mathcal{V}_m \subset \mathcal{V}$. Then $f \notin \mathcal{N}_{m,k}$, so that $f \in \mathcal{H}_{m,k}$. Hence, f has a periodic orbit, say Q_m, in $\mathcal{V}_m$ with index k by definition of $\mathcal{H}_{m,k}$. This completes the proof. □

In the following lemma, we show that if Λ is a chain transitive set of a diffeomorphism $f \in \mathcal{R}_1$ and $f|_\Lambda$ is shadowing, then every periodic point in Λ has the same index.

Lemma 4.3.2 *Let $f \in \mathcal{R}_1$ and let Λ be a chain transitive set of f. If $f|_\Lambda$ is shadowing, then all periodic points in Λ have the same index.*

Proof Consider periodic points $p, q \in \mathrm{Per}(f) \cap \Lambda$ and let $\varepsilon > 0$ be small enough so that the local stable manifold $W^s_\varepsilon(p)$ and the local unstable manifold $W^u_\varepsilon(q)$ of size ε are well defined.

Take $d > 0$ such that every d-pseudotrajectory in Λ is ε-shadowed by a point in M.

Since Λ is chain transitive, there is a finite d-pseudotrajectory $\{x_0, x_1, \dots, x_n\}$ of f in Λ such that $x_0 = q$ and $x_n = p$.

Construct a d-pseudotrajectory ξ in Λ as follows:

$$\xi = \{\dots, f^{-2}(q), f^{-1}(q), q, x_1, x_2, \dots, p, f(p), f^2(p), \dots\}.$$

Then there is an orbit $O(y,f)$ that ε-shadows ξ.

Since $\mathrm{Orb}(y) \cap W^s_\varepsilon(p) \neq \emptyset$ and $\mathrm{Orb}(y) \cap W^u_\varepsilon(q) \neq \emptyset$, we have the inclusion $y \in W^s(p) \cap W^u(q)$.

This implies that the indices of p and of q are the same. Indeed, since $W^s(p) \cap W^u(q) \neq \emptyset$ and $W^s(q) \cap W^u(p) \neq \emptyset$, the transversality of the intersections implies that

$$\dim W^s(p) + \dim W^u(q) \geq \dim M \quad \text{and} \quad \dim W^s(q) + \dim W^u(p) \geq \dim M.$$

These inequalities imply that $\dim W^s(p) = \dim W^s(q)$. Indeed, it follows from the inequality

$$\dim W^s(p) \geq \dim M - \dim W^u(q) = \dim W^s(q)$$

that $\dim W^s(p) \geq \dim W^s(q)$. A similar reasoning shows that $\dim W^s(q) \geq \dim W^s(p)$. This completes the proof. □

Proof Now we define the residual subset $\mathscr{R} \subset \mathrm{Diff}^1(M)$ for which the assertion of Theorem 4.3.1 holds as follows:

$$\mathscr{R} = \mathscr{R}_1 \cap \mathscr{R}_2.$$

The following proposition is crucial for the proof of Theorem 4.3.1.

Proposition 4.3.1 *Let $f \in \mathscr{R}$ and let Λ be a chain transitive set of f that is locally maximal. Then there exist constants $m > 0$ and $0 < \lambda < 1$ such that for any $p \in Per(f) \cap \Lambda$,*

$$\prod_{i=0}^{\pi(p)-1} \left\| Df^m|_{E^s(f^{im}(p))} \right\| < \lambda^{\pi(p)},$$

$$\prod_{i=0}^{\pi(p)-1} \left\| Df^{-m}|_{E^u(f^{-im}(p))} \right\| < \lambda^{\pi(p)},$$

and

$$\left\| Df^m|_{E^s(p)} \right\| \cdot \left\| Df^{-m}|_{E^u(f^m(p))} \right\| < \lambda^2.$$

Proof Since $f \in \mathscr{R}_1$, periodic points of f are hyperbolic and dense in Λ. By Lemma 4.3.2, they have the same index.

First we show that there exists a C^1 neighborhood $\mathscr{U}(f)$ of $f \in \mathscr{R}_2$ and a neighborhood U of Λ such that every $g \in \mathscr{U}(f)$ does not have nonhyperbolic periodic orbits contained in U.

To get a contradiction, assume that for any C^1 neighborhood $\mathscr{V}(f)$ of f and a neighborhood V of Λ, there is a diffeomorphism $g \in \mathscr{V}(f)$ having a nonhyperbolic periodic orbit Q in V.

Applying a C^1-small perturbation of the diffeomorphism g, we can assume that there are diffeomorphisms $g_1, g_2 \in \mathscr{V}(f)$ and hyperbolic periodic orbits Q_1 and Q_2 in V of g_1 and g_2, respectively, such that index $Q_1 \neq$ index Q_2.

Indeed, assume that Q is nonhyperbolic and take a point $q \in Q$. Let $l > 0$ be the period of Q. Then $Dg^l(q)$ has an eigenvalue with absolute value equal to one.

Applying the Franks lemma (Lemma 3.2.1), we can find a C^1-small perturbation of g (denoted again g) such that there is a g^l-invariant small arc $\mathscr{I}$ centered at q and such that $g^l|_{\mathscr{I}}$ is the identity map.

Applying additional C^1-small perturbations of g, we can construct diffeomorphisms g_1 and g_2 C^1-close to g and having hyperbolic periodic orbits Q_1 and Q_2 in V with different indices.

Hence, we can construct two sequences of diffeomorphisms g_n and g'_n that converge to f in $\mathrm{Diff}^1(M)$ and two sequences of hyperbolic periodic orbits Q_n, Q'_n of g_n and g'_n, respectively, such that

$$\lim_{n\to\infty} Q_n = \tilde{\Lambda} = \lim_{n\to\infty} Q'_n$$

and index $Q_n \neq$ index Q'_n for each $n \in \mathbb{N}$.

Without loss of generality, taking a subsequence if necessary, we may assume that index $Q_n =$ index Q_m and index $Q'_n =$ index Q'_m for all $m, n \in \mathbb{N}$.

Applying Lemma 4.3.1 to the f-invariant set $\tilde{\Lambda}$, we can choose two sequences of periodic orbits P_n and P'_n of f such that index P_n = index Q_n, index P'_n = index Q'_n, and $\tilde{\Lambda}$ is the Hausdorff limit of $\{P_n\}$ and $\{P'_n\}$, respectively. Since Λ is locally maximal and $\tilde{\Lambda} \subset \Lambda$, we may assume that $P_n, P'_n \subset \Lambda$ for sufficiently large n. Since index $P_n \neq$ index P'_n, we get a contradiction with Lemma 4.3.2.

Note that the reasoning used in the above proof shows that all of the indices of periodic orbits of $g \in \mathscr{U}(f)$ contained in U are the same. Hence, by the reason stated in the paragraph located before Proposition 4.2.1 and Propositions 4.2.1–4.2.2, we get constants $K > 0$, $m_0 \in \mathbb{Z}^+$, and $0 < \lambda < 1$ such that for any periodic point $p \in \Lambda$ with $\pi(p) \geq K$, the following inequalities hold:

$$\prod_{i=0}^{\pi(p)-1} \left\| Df^{m_0}|_{E^s(f^{im_0}(p))} \right\| < \lambda^{\pi(p)},$$

$$\prod_{i=0}^{\pi(p)-1} \left\| Df^{-m_0}|_{E^u(f^{-im_0}(p))} \right\| < \lambda^{\pi(p)},$$

and

$$\left\| Df^{m_0}|_{E^s(p)} \right\| \cdot \left\| Df^{-m_0}|_{E^u(f^{m_0}(p))} \right\| < \lambda^2.$$

Let Λ_0 be the set of all periodic points in Λ whose periods are less than K. Since every periodic point of f is hyperbolic, Λ_0 is a finite set; hence, Λ_0 is a hyperbolic set of f.

Let k be a positive integer such that

$$\left\| Df^{km_0}|_{E^s(x)} \right\| < \lambda \text{ and } \left\| Df^{km_0}|_{E^u(x)} \right\| < \lambda$$

for all $x \in \Lambda_0$. If we take $m = km_0$, then it is easy to show that m and λ are the required constants. □

Now we complete the proof of Theorem 4.3.1.

Let Λ be a locally maximal chain transitive set of $f \in \mathcal{R}$. To get the conclusion, it is enough to show that if $f|_\Lambda$ is shadowing, then Λ is hyperbolic.

Now let us check that f^m satisfies all the assumptions (P.1)–(P.4) of Proposition 3.2.4.

Let U be an isolating block of Λ so that $\Lambda_f(U) = \Lambda$. By the third property of Proposition 4.3.1, we can see that Λ admits a dominated splitting $T_\Lambda M = E \oplus F$ for f^m that satisfies $E(p) = E^s(p)$ and $F(p) = E^u(p)$ for every $p \in \mathrm{Per}(f) \cap \Lambda$. In fact, it has shown in the proof of Proposition 4.3.1 that there are a C^1 neighborhood $\mathcal{U}(f)$ of $f \in \mathcal{R}_2$ and a neighborhood U of Λ such that every $g \in \mathcal{U}(f)$ does not have nonhyperbolic periodic orbits contained in U. Thus, the assertion follows from Proposition 4.2.1 (note that by Lemmas 4.3.1 and 4.3.2, $\Lambda = \overline{P_{\dim E}(f)}$). Since $f|_\Lambda$ is shadowing, $f^m|_\Lambda$ is also shadowing and thus, (P.1)–(P.4) are satisfied for f^m.

Therefore, Λ is hyperbolic for f^m by Proposition 3.2.4, so that Λ is hyperbolic for f as well. □

Historical Remarks Theorem 4.3.1 was first proved by K. Lee and X. Wen in [34] with application of the Mañé ergodic closing lemma [42]. We prove the result applying Proposition 3.2.4 (and do not use the ergodic closing lemma). The proof is a little bit longer, but it is simplified making an effective use of the shadowing property.

For expansive homoclinic classes, a similar result to Theorem 4.3.1 was proved by Yang-Gan [110] without the local maximality assumption of the sets.

References

1. Abdenur, F., Diaz, L.J.: Pseudo-orbit shadowing in the C^1 topology. Discrete Contin. Dyn. Syst. **17**, 223–245 (2007)
2. Andronov, A., Pontryagin, L.: Systèmes grossiers. Dokl. Akad. Nauk SSSR **14**, 247–250 (1937)
3. Anosov, D.V.: Geodesic Flows on Closed Riemannian Manifolds of Negative Curvature. Amer. Math. Soc., Providence, RI (1969)
4. Anosov, D.V.: On a class of invariant sets of smooth dynamical systems. In: Proc. 5th Int. Conf. Nonl. Oscill., vol. 2, pp. 39–45, Kiev (1970)
5. Aoki, N.: Topological Dynamics. Topics on General Topology, pp. 625–740. North-Holland, Amsterdam (1989)
6. Aoki, N., Hiraide, K.: Topological Theory of Dynamical Systems. Recent Advances. North-Holland Math. Library, vol. 52. North-Holland, Amsterdam (1994)
7. Aoki, N.: The set of Axiom A diffeomorphisms with no cycle. Bol. Soc. Brasil Mat. (NS) **23**, 21–65 (1992)
8. Arnold, V.I.: Ordinary Differential Equations. Universitext. Springer, Berlin (2006)
9. Birkhoff, G.: An extension of Poincaré's last geometric theorem. Acta Math. **47**, 297–311 (1926)
10. Birkhoff, G.: Dynamical Systems. Colloquium Publ., vol. 9. Amer. Math. Soc., New York (1927)
11. Bonatti, C., Díaz, L.J., Viana, M.: Dynamics Beyond Uniform Hyperbolicity. Encyclopedia of Mathematical Sciences (Mathematical Physics), vol. 102. Springer, Berlin (2005)
12. Bowen, R.: Equilibrium States and the Ergodic Theory of Anosov Diffeomorphisms. Lect. Notes Math., vol. 470. Springer, Berlin (1975)
13. Coppel, W.A.: Dichotomies in Stability Theory. Lect. Notes Math., vol. 629. Springer, Berlin (1978)
14. Conley, C.: The Gradient Structure of a Flow. IBM Research RC 3932 (17806), Yorktown Heights, NY (1972)
15. Conley, C.: Isolated Invariant Sets and Morse Index. CBMS Regional Conferences Series in Math., vol. 38. Amer. Math. Soc., Providence, RI (1978)
16. Crovisier, S.: Periodic orbits and chain-transitive sets of C^1-diffeomorphisms. IHÉS Publ. Math. **104**, 87–141 (2006)
17. Ding, H.: Disturbance of the homoclinic trajectory and applications. Acta Sci. Nat. Univ. Pekin. **1**, 53–63 (1986)
18. Franke, J.E., Selgrade, J.F.: Hyperbolicity and chain recurrence. J. Differ. Equ. **26** (1977), 27–36.

S.Yu. Pilyugin, K. Sakai, *Shadowing and Hyperbolicity*, Lecture Notes in Mathematics 2193, DOI 10.1007/978-3-319-65184-2

19. Franks, J.: Necessary conditions for stability of diffeomorphisms. Trans. Am. Math. Soc. **158**, 301–308 (1971)
20. Gan, S.: A generalized shadowing lemma. Discrete Contin. Dyn. Syst. **8**, 627–632 (2002)
21. Gan, S., Wen, L.: Nonsingular star flows satisfy Axiom A and the no-cycle condition. Invent. Math. **164**, 279–315 (2006)
22. Guchenheimer, J.: A Strange, Strange Attractor. The Hopf Bifurcation Theorem and Its Applications. Applied Mathematical Series, vol. 19, pp. 368–381. Springer, New York (1976)
23. Hammel, S.M., Yorke, J.A., Grebogi, C.: Do numerical orbits of chaotic dynamical processes represent true orbits? J. Complexity **3**, 136–145 (1987)
24. Hammel, S.M., Yorke, J.A., Grebogi, C.: Numerical orbits of chaotic processes represent true orbits. Bull. Am. Math. Soc. (N.S.) **19**, 465–469 (1988)
25. Hayashi, S.: Diffeomorphisms in $\mathscr{F}1(M)$ satisfy Axiom A. Ergod. Theory Dyn. Syst. **12**, 233–253 (1992)
26. Hayashi, S.: Connecting invariant manifolds and the solution of the C^1 stability and Ω-stability conjecture. Ann. Math. **145**, 81–137 (1997)
27. Hirsch, M., Pugh, C., Shub, M.: Invariant Manifolds. Lect. Notes Math., vol. 583. Springer, Berlin (1977)
28. Katok, A., Hasselblatt, B.: Introduction to the Modern Theory of Dynamical Systems. Encyclopedia of Mathematics and Its Applications, vol. 54. Cambridge Univ. Press, Cambridge (1995)
29. Komuro, M.: One-parameter flows with the pseudo orbit tracing property. Monatsh. Math. **98**, 219–253 (1984)
30. Komuro, M.: Lorenz attractors do not have the pseudo-orbit tracing property. J. Math. Soc. Jpn. **37**, 489–514 (1985)
31. Kupka, I.: Contributions à la théorie des champs génériques. Contrib. Diff. Equ. **2**, 457–484 (1963)
32. Kuratowskii, C.: Topology II. Academic Press, New York, London (1968)
33. Lee, K., Sakai, K.: Structural stability of vector fields with shadowing. J. Differ. Equ. **232**, 303–313 (2007)
34. Lee, K., Wen, X.: Shadowable chain transitive sets of C^1-generic diffeomorphisms. Bull. Korean Math. Soc. **49**, 263–270 (2012)
35. Lee, K., Moriyasu, K. Sakai, K.: C^1-stable shadowing diffeomorphisms. Discrete Contin. Dyn. Syst. **22**, 683–697 (2008)
36. Liao, S.T.: The Qualitative Theory of Differentiable Dynamical Systems. Science Press, Beijing (1996)
37. Li, C., Wen, L.: $\mathscr{X}^*$ plus Axiom A does not imply no-cycle. J. Differ. Equ. **119**, 395–400 (1995)
38. Maizel, A.D.: On stability of solutions of systems of differential equations, Trudy Ural. Politehn. Inst. **51**, 20–50 (1954)
39. Mañé, R.: Characterizations of AS Diffeomorphisms. Geometry and Topology. Lect. Notes Math., vol. 597, pp. 389–394. Springer, Berlin (1977)
40. Mañé, R.: Contributions to stability conjecture. Topology **17**, 383–396 (1978)
41. Mañé, R.: Persistent manifolds are normally hyperbolic. Trans. Am. Math. Soc. **246**, 261–283 (1978)
42. Mañé, R.: An ergodic closing lemma. Ann. Math. **116**, 503–540 (1982)
43. Mañé, R.: Ergodic Theory and Differentiable Dynamics. Ergebnisse der Mathematik und ihrer Grenzgebiete, vol. 8. Springer, Berlin (1987)
44. Mañé, R.: On the creation of homoclinic points. IHÉS Publ. Math. **66**, 139–159 (1988)
45. Mañé, R.: A proof of the C^1-stability conjecture. IHÉS Publ. Math. **66**, 161–210 (1988)
46. Morimoto, A.: The Method of Pseudo-orbit Tracing and Stability of Dynamical Systems. Sem. Note, vol. 39. Tokyo Univ. (1979)
47. Moriyasu, K.: The topological stability of diffeomorphisms. Nagoya Math. J. **123**, 91–102 (1991)

48. Newhouse, S.: Hyperbolic limit sets. Trans. Am. Math. Soc. **167**, 125–150 (1972)
49. Ombach, J.: Shadowing, expansiveness and hyperbolic homeomorphisms. J. Aust. Math. Soc. (Ser. A) **61**, 57 – 72 (1996).
50. Osipov, A.V., Pilyugin, S.Yu., Tikhomirov, S.B.: Periodic shadowing and Ω-stability. Regul. Chaotic Dyn. **15**, 404–417 (2010)
51. Pacifico, M.J., Pujals, E.R., Vieitez, J.L.: Robustly expansive homoclinic classes. Ergod. Theory Dyn. Syst. **25**, 271–300 (2005)
52. Palis, J., Smale, S.: Structural Stability Theorems. Global Analysis, Symp. Pure Math., vol. 14, pp. 223–231. Amer. Math. Soc., New York (1970)
53. Palis, J.: On the C^1 Ω-stability conjecture. IHÉS Publ. Math. **66**, 211–215 (1988)
54. Palis, J., Takens, F.: Hyperbolicity and Sensitive Chaotic Dynamics at Homoclinic Bifurcations. Cambridge Studies in Advanced Mathematics, vol. 35. Cambridge University Press, Cambridge (1993)
55. Palmer, K.: Exponential dichotomies and Fredholm operators. Proc. Am. Math. Soc. **104**, 149–156 (1988)
56. Palmer, K.: Shadowing in Dynamical Systems. Theory and Applications. Kluwer, Dordrecht (2000)
57. Palmer, K.J., Pilyugin, S.Yu., Tikhomirov, S.B.: Lipschitz shadowing and structural stability of flows. J. Differ. Equ. **252**, 1723–1747 (2012)
58. Petrov, A.A., Pilyugin, S.Yu.: Shadowing near nonhyperbolic fixed points. Discrete Contin. Dyn. Syst. **34**, 3761–3772 (2014)
59. Petrov, A.A., Pilyugin, S.Yu.: Nonsmooth mappings with Lipschitz shadowing. arXiv:1510.03074 [math.DS] (2015)
60. Pilyugin, S.Yu.: Introduction to Structurally Stable Systems of Differential Equations. Birkhauser-Verlag, Basel (1992)
61. Pilyugin, S.Yu.: The Space of Dynamical Systems with the C^0 Topology. Lect. Notes Math., vol. 1571. Springer, Berlin (1994)
62. Pilyugin, S.Yu.: Shadowing in structurally stable flows. J. Differ. Equ. **140**, 238–265 (1997)
63. Pilyugin, S.Yu., Plamenevskaya, O.B.: Shadowing is generic. Topology Appl. **97**, 253–266 (1999)
64. Pilyugin, S.Yu.: Shadowing in Dynamical Systems. Lect. Notes Math., vol. 1706. Springer, Berlin (1999)
65. Pilyugin, S.Yu., Rodionova A.A., Sakai K.: Orbital and weak shadowing properties. Discrete Contin. Dyn. Syst. **9**, 404–417 (2003)
66. Pilyugin, S.Yu., Sakai, K.: C^0 transversality and shadowing properties. Proc. Steklov Inst. Math. **256**, 290–305 (2007)
67. Pilyugin, S.Yu.: Variational shadowing. Discrete Contin. Dyn. Syst., Ser. B. **14**, 733–737 (2010)
68. Pilyugin, S.Yu., Tikhomirov, S.B.: Lipschitz shadowing implies structural stability. Nonlinearity **23**, 2509–2515 (2010)
69. Pilyugin, S.Yu., Tikhomirov, S.B.: Vector fields with the oriented shadowing property. J. Differ. Equ. **248**, 1345–1375 (2010)
70. Pilyugin, S.Yu.: Theory of pseudo-orbit shadowing in dynamical systems. Differ. Equ. **47**, 1929–1938 (2011)
71. Pilyugin, S.Yu.: Spaces of Dynamical Systems. Walter de Gruyter, Berlin/Boston (2012)
72. Plamenevskaya, O.B.: Weak shadowing for two-dimensional diffeomorphisms. Vestnik S.Petersb. Univ., Math. **31**, 49–56 (1999)
73. Pliss, V.A.: On a conjecture due to Smale. Diff. Uravnenija **8**, 268–282 (1972)
74. Pliss, V.A.: Bounded Solutions of Nonhomogeneous Linear Systems of Differential Equations. Problems in the Asymptotic Theory of Nonlinear Oscillations, pp. 168–173, Kiev (1977)
75. Poincaré, H.: Les Méthodés Nouvelles de la Mécanique Céleste. Paris (1892–1899)
76. Pugh, C., Shub, M.: Ω-stability theorem for vector fields. Invent. Math. **11**, 150–158 (1970)

77. Pugh, C., Robinson, C.: The C^1 closing lemma, including Hamiltonians. Ergod. Theory Dyn. Syst. **3**, 261–313 (1983)
78. Robbin, J.: A structural stability theorem. Ann. Math. **94**, 447–493 (1971)
79. Robinson, C.: Structural stability of vector fields. Ann. Math. **99**, 154–175 (1974)
80. Robinson, C.: Structural stability of C^1 flows. In: Proc. Symp. Dyn. Syst. (University of Warwick, 1974). Lect. Notes Math., vol. 468. Springer, New York (1975)
81. Robinson, C.: Structural stability for C^1-diffeomorphisms. J. Differ. Equ. **22**, 28–73 (1976)
82. Robinson, C.: C^r structural stability implies Kupka–Smale. In: Peixoto, M. (ed.) Dynamical Systems, pp. 443–449. Acad. Press, New York (1976)
83. Robinson, C.: Stability theorems and hyperbolicity in dynamical systems. Rocky Mount. J. Math. **7**, 425–437 (1977)
84. Robinson, C.: Dynamical Systems: Stability, Symbolic Dynamics, and Chaos, 2nd edn. Studies in Advanced Mathematics. CRC Press, Boca Raton (1999)
85. Sacker, R.J., Sell, G.R.: Existence of dichotomies and invariant splittings for linear differential systems. II. J. Differ. Equ. **22**, 478–496 (1976)
86. Sacker, R.J., Sell, G.R.: A spectral theory for linear differential systems. J. Differ. Equ. **27**, 320–358 (1978)
87. Sakai, K.: Pseudo-orbit tracing property and strong transversality of diffeomorphisms on closed manifolds. Osaka J. Math. **31**, 373–386 (1994)
88. Sakai, K.: Shadowing Property and Transversality Condition. Dynamical Systems and Chaos, vol. 1, pp. 233-238. World Scientific, Singapore (1995)
89. Sakai, K. Diffeomorphisms with weak shadowing. Fund. Math. **168**, 57–75 (2001)
90. Sakai, K.: C^1-stably shadowable chain components. Ergod. Theory Dyn. Syst. **28**, 987–1029 (2008)
91. Sakai, K.: Shadowable chain transitive sets. J. Differ. Equ. Appl. **19**, 1601–1618 (2013)
92. Sawada, K.: Extended f-orbits are approximated by orbits. Nagoya Math. J. **79** 33–45 (1980)
93. Shimomura, T.: On a structure of discrete dynamical systems from the viewpoint of chain components and some applications. Jpn. J. Math. **15**, 100–126 (1989)
94. Smale, S.: Stable manifolds for differential equations and diffeomorphisms. Ann. Schuola Norm. Sup. Pisa **17**, 97–116 (1963)
95. Smale, S.: Differentiable dynamical systems. Bull. Am. Math. Soc. **73**, 747–817 (1967)
96. Smale, S.: Structurally stable diffeomorphisms with infinitely many periodic points. In: Proc. Intern. Conf. Nonl. Oscill., vol. 2, pp. 365–366, Kiev (1963)
97. Smale, S.: Diffeomorphisms with Many Periodic Points. Differential and Combinatorial Topology, pp. 63–80. Princeton University Press, Princeton (1965)
98. Smale, S.: The Ω-stability theorem. In: Global Analysis, Symp. Pure Math., vol. 14, pp. 289–297. Amer. Math. Soc., New York (1970)
99. Tikhomirov, S.B.: Interiors of sets of vector fields with shadowing properties corresponding to some classes of reparametrizations. Vestn. St.Petersb. Univ. Ser. 1 **41**, 360–366 (2008)
100. Tikhomirov, S.B.: An example of a vector field with the oriented shadowing property. J. Dyn. Control Syst. **21**, 643–654 (2015)
101. Tikhomirov, S.B.: Holder shadowing on finite intervals. Ergod. Theory Dyn. Syst. **35**, 2000–2016 (2015)
102. Thomas, R.F.: Stability properties of one-parameter flows. Proc. Lond. Math. Soc. **45**, 479–505 (1982)
103. Toyoshiba, H.: Vector fields in the interior of Kupka–Smale systems satisfy Axiom A. J. Differ. Equ. **177**, 27–48 (2001)
104. Walters, P.: Anosov diffeomorphisms are topologically stable. Topology **9**, 71–78 (1970)
105. Walters, P.: On the Pseudo Orbit Tracing Property and Its Relationship to Stability. The Structure of Attractors in Dynamical Systems. Lect. Notes Math., vol. 668, pp. 231–244. Springer, Berlin (1978)
106. Wen, L.: On the C^1 stability conjecture for flows. J. Differ. Equ. **129**, 334–357 (1996)
107. Wen, L.: The selecting lemma of Liao. Discrete Contin. Dyn. Syst. **20**, 159–175 (2008)

108. Wen, L.: Differentiable Dynamical Systems. An Introduction to Structural Stability and Hyperbolicity. Graduate Studies in Mathematics, vol. 173. Amer. Math. Soc., Providence, RI (2016)
109. Wen, X., Gan, S., Wen, L.: C^1-stably shadowable chain components are hyperbolic. J. Differ. Equ. **246**, 340–357 (2009)
110. Yang, D., Gan, S.: Expansive homoclinic classes. Nonlinearity **22**, 729–733 (2009)

Index

C^0 topology, 13
C^1 topology, 14, 21
Ω-equivalent flows, 22
Ω-stable diffeomorphism, 15
Ω stable vector field, 22
λ-hyperbolic pair of sequences, 151
λ-quasi-hyperbolic δ-pseudotrajectory, 147
λ-quasi-hyperbolic pair of sequences, 151
λ-quasi-hyperbolic string, 136
π-invariant subbundle, 51
ε-chain, 5
ε-shadowed pseudotrajectory, 2
d-orbit, 2
d-pseudotrajectory, 2, 9
k-cycle, 17
1-cycle, 16

Lipschitz shadowing property on a set, 75

adapted (Lyapunov) norm, 27
analytic strong transversality condition, 18

balance sequence, 151
balance sequence adapted to a λ-quasi-hyperbolic sequence pair, 151
basic set, 16, 22, 187

chain equivalent point, 6
chain recurrence class, 6
chain recurrent point, 5, 23
chain recurrent set, 6, 24
chain transitive set, 6, 182
class $\mathscr{B}$, 155
Condition B, 56
Condition S, 68
continuation of an invariant set, 187

discrete Lipschitz shadowing property, 110
dominated splitting, 18

expansive homeomorphism, 64
expansivity constant, 64

finite Lipschitz shadowing property, 4
finite shadowing property, 3

generic property, 18
geometric strong transversality condition, 17, 22

Hölder shadowing property on finite intervals, 5
homoclinic class, 182
homoclinic point, 19
homoclinic point of the closed trajectory, 25
homoclinic point of the rest point, 24
homoclinically related periodic points, 182
hyperbolic periodic point, 17
hyperbolic rest point, 23
hyperbolic sequence, 38
hyperbolic set, 12, 19

S.Yu. Pilyugin, K. Sakai, *Shadowing and Hyperbolicity*, Lecture Notes in Mathematics 2193, DOI 10.1007/978-3-319-65184-2

hyperbolic structure, 12
hyperbolicity constants, 12

isolating block, 187

Kupka–Smale diffeomorphism, 17
Kupka–Smale flow, 23

Lipschitz periodic shadowing property, 4
Lipschitz shadowing property, 4, 10
locally maximal set, 187

maximal chain transitive set, 6

no cycle condition, 17
nonwandering point, 13
nonwandering set, 13

orbital shadowing property, 3, 11
oriented shadowing property, 11

periodic shadowing property, 4
Perron property, 38
pseudoorbit tracing property, 3

reparametrization, 10
residual set, 18

shadowing property, 2
shadowing trajectory, 2
stable manifold for a diffeomorphism, 12
stable manifold for a flow, 20
standard shadowing property, 3, 10, 181
star systems, 17
structurally stable diffeomorphism, 15
structurally stable vector field, 21

topologically equivalent flows, 21
topologically stable diffeomorphism, 131
transverse homoclinic point, 19, 25

unstable manifold for a diffeomorphism, 12
unstable manifold for a flow, 20

weak shadowing property, 132
well adapted balance sequence, 151

LECTURE NOTES IN MATHEMATICS

Editorial Policy

1. Lecture Notes aim to report new developments in all areas of mathematics and their applications – quickly, informally and at a high level. Mathematical texts analysing new developments in modelling and numerical simulation are welcome.

   Manuscripts should be reasonably self-contained and rounded off. Thus they may, and often will, present not only results of the author but also related work by other people. They may be based on specialised lecture courses. Furthermore, the manuscripts should provide sufficient motivation, examples and applications. This clearly distinguishes Lecture Notes from journal articles or technical reports which normally are very concise. Articles intended for a journal but too long to be accepted by most journals, usually do not have this "lecture notes" character. For similar reasons it is unusual for doctoral theses to be accepted for the Lecture Notes series, though habilitation theses may be appropriate.

2. Besides monographs, multi-author manuscripts resulting from SUMMER SCHOOLS or similar INTENSIVE COURSES are welcome, provided their objective was held to present an active mathematical topic to an audience at the beginning or intermediate graduate level (a list of participants should be provided).

   The resulting manuscript should not be just a collection of course notes, but should require advance planning and coordination among the main lecturers. The subject matter should dictate the structure of the book. This structure should be motivated and explained in a scientific introduction, and the notation, references, index and formulation of results should be, if possible, unified by the editors. Each contribution should have an abstract and an introduction referring to the other contributions. In other words, more preparatory work must go into a multi-authored volume than simply assembling a disparate collection of papers, communicated at the event.

3. Manuscripts should be submitted either online at www.editorialmanager.com/lnm to Springer's mathematics editorial in Heidelberg, or electronically to one of the series editors. Authors should be aware that incomplete or insufficiently close-to-final manuscripts almost always result in longer refereeing times and nevertheless unclear referees' recommendations, making further refereeing of a final draft necessary. The strict minimum amount of material that will be considered should include a detailed outline describing the planned contents of each chapter, a bibliography and several sample chapters. Parallel submission of a manuscript to another publisher while under consideration for LNM is not acceptable and can lead to rejection.

4. In general, **monographs** will be sent out to at least 2 external referees for evaluation.

   A final decision to publish can be made only on the basis of the complete manuscript, however a refereeing process leading to a preliminary decision can be based on a pre-final or incomplete manuscript.

   Volume Editors of **multi-author works** are expected to arrange for the refereeing, to the usual scientific standards, of the individual contributions. If the resulting reports can be forwarded to the LNM Editorial Board, this is very helpful. If no reports are forwarded or if other questions remain unclear in respect of homogeneity etc, the series editors may wish to consult external referees for an overall evaluation of the volume.

5. Manuscripts should in general be submitted in English. Final manuscripts should contain at least 100 pages of mathematical text and should always include

   – a table of contents;
   – an informative introduction, with adequate motivation and perhaps some historical remarks: it should be accessible to a reader not intimately familiar with the topic treated;
   – a subject index: as a rule this is genuinely helpful for the reader.
   – For evaluation purposes, manuscripts should be submitted as pdf files.

6. Careful preparation of the manuscripts will help keep production time short besides ensuring satisfactory appearance of the finished book in print and online. After acceptance of the manuscript authors will be asked to prepare the final LaTeX source files (see LaTeX templates online: https://www.springer.com/gb/authors-editors/book-authors-editors/manuscriptpreparation/5636) plus the corresponding pdf- or zipped ps-file. The LaTeX source files are essential for producing the full-text online version of the book, see http://link.springer.com/bookseries/304 for the existing online volumes of LNM). The technical production of a Lecture Notes volume takes approximately 12 weeks. Additional instructions, if necessary, are available on request from lnm@springer.com.

7. Authors receive a total of 30 free copies of their volume and free access to their book on SpringerLink, but no royalties. They are entitled to a discount of 33.3 % on the price of Springer books purchased for their personal use, if ordering directly from Springer.

8. Commitment to publish is made by a *Publishing Agreement*; contributing authors of multiauthor books are requested to sign a *Consent to Publish form*. Springer-Verlag registers the copyright for each volume. Authors are free to reuse material contained in their LNM volumes in later publications: a brief written (or e-mail) request for formal permission is sufficient.

Addresses:

Professor Jean-Michel Morel, CMLA, École Normale Supérieure de Cachan, France
E-mail: moreljeanmichel@gmail.com

Professor Bernard Teissier, Equipe Géométrie et Dynamique,
Institut de Mathématiques de Jussieu – Paris Rive Gauche, Paris, France
E-mail: bernard.teissier@imj-prg.fr

Springer: Ute McCrory, Mathematics, Heidelberg, Germany,
E-mail: lnm@springer.com

Printed in the United States
By Bookmasters

Printed in the United States
By Bookmasters